Dynamical Systems and Differential Geometry via MAPLE

Dynamical Systems and Differential Geometry via MAPLE

By

Constantin Udriste and Ionel Tevy

Cambridge
Scholars
Publishing

Dynamical Systems and Differential Geometry via MAPLE

By Constantin Udriste and Ionel Tevy

This book first published 2021

Cambridge Scholars Publishing

Lady Stephenson Library, Newcastle upon Tyne, NE6 2PA, UK

British Library Cataloguing in Publication Data
A catalogue record for this book is available from the British Library

ISBN (10): 1-5275-7223-4
ISBN (13): 978-1-5275-7223-2

Contents

Preface

Our book "Dynamical Systems and Differential Geometry via MAPLE" is designed starting from the course of the same name taught (by the first author) to engineering students, year III, from the Faculty of Applied Sciences, the Mathematical-Computer Section, in the University Politehnica of Bucharest, since 2005. It condenses the offer and the importance of the mutual impact and interconditioning of the three mathematical disciplines. Some subjects are taught at other faculties or universities in the world, especially in the master or doctorate courses, being un-missed from the papers that can be published in journals now categorized as good or very good.

The subject of the course includes twelve Chapters: Dynamical systems; Invariant sets, conservation and dissipation; The stability of equilibrium positions and Hopf's bifurcation; Gyroscopic forces and collision avoidance; Nambu dynamical systems; Geometric dynamics and wind theory; Motions of a curve; Motions of a surface; Evolutive Riemannian metrics; Evolution of Tzitzeica hypersurfaces; Dynamic of evolutive optimization problems; Tasks with Maple. The problems (self-evaluation or research) in each Chapter, while not at all trivial, tremendously enhance one's understanding of the material.

All Chapters were structured by importance, accessibility and impact of theoretical notions capable of shaping a future mathematician-computer scientist possessing knowledge of evolutionary dynamical systems. The intermediate variants of the manuscript have extended over the years, leading to the selection of the most important and manageable notions and reaching maturity through this version of the book that we have decided to publish.

The scientific authority we have has allowed us to impose the point of view on the thematic and on the types of rations that deserve to be offered to readers of texts of dynamical systems, differential geometry and Maple. Everything was structured for the benefit of optimizing evolutionary geometric aspects that describe significant physical or engineering phenomena. It is known that to avoid blocking the students' minds, a simplified language is preferred, specific to the applied mathematics, as we have done as a rule, sending the pure mathematics readers to bibliography. Our long experience as a mathematics professors at a technical university has helped us to fluently expose ideas stripped of excessive formalization, with great impact on students' understanding of natural phenomena encountered in engineering and economics as well as their dressing in mathematical clothes for the intellectual holiday. In this sense, we preferred evolutionary models that intelligently present the real world and we have avoided the hashish of abstract notions specific to theorists.

The novelty problems were finalized following repeated discussions with the mathematics and physics professors from The Faculty of Applied Sciences, especially with those interested in dynamical systems and differential geometry, with Maple simulations, to whom we thank and thoughts of appreciation. We are open to any kind of remarks or criticisms that bring didactic benefits to this course of dynamic evolutionary systems.

The area of Dynamical Systems and Differential Geometry via MAPLE is a mixed field which has become exceedingly technical in recent years. Everything was structured for the benefit of optimizing evolutionary geometric aspects that describe significant physical or engineering phenomena. Also, Maple has a large number of subroutine libraries which have been expressly written to provide solutions to standard homework problems for students.

With this book, the authors provide a self-contained and accessible introduction for graduate or advanced undergraduate students in mathematics, engineering, physics, and economic sciences. Both classical and modern methods used in the field are described in detail, concentrating on the model cases that simplify the presentation without compromising the deep technical aspects of the theory, thus allowing students to learn the material in a short period of time. This book is suitable for self-study for students and professors with a background in differential geometry both, and for teaching a semester-long introductory graduate course in Dynamical Systems and Differential Geometry via MAPLE. Copious exercises are included in each Chapter, and applications of the theory are also presented to connect dynamical systems and differential geometry with the more general areas of dynamical systems, differential geometry and mathematical physics, in large.

The mottos used at Chapters are stanzas from George Cosbuc's poems (in Romanian), selected by the authors and translated in English by Associate Prof. Dr. Brandusa Prepelita-Raileanu, University Politehnica of Bucharest. George Cosbuc (20 September 1866 - 9 May 1918) was a Romanian poet, translator, teacher, and journalist, best remembered for his verses describing, praising and eulogizing rural life, its many travails but also its occasions for joy.

The authors thank Miss Associate Prof. Dr. Oana-Maria Pastae, "Constantin Brancusi" University of Tg-Jiu, for the English improvement of the manuscript.

February 2021

Prof. Emeritus Dr. Constantin Udriste[1], Prof. Dr. Ionel Tevy

Department of Mathematics and Informatics, Faculty of Applied Sciences, University Politehnica of Bucharest, Splaiul Independentei 313, RO-060042, Bucharest, 6, Romania; udriste@mathem.pub.ro ; vascatevy@yahoo.fr

[1] Academy of Romanian Scientists, Ilfov 3, RO-050044 Bucharest 5, Romania.

Chapter 1

DYNAMICAL SYSTEMS

Motto:
Some people hold dear
What's vain in the other men's eye
But He who can scan both the earth and the sky
And set up a bridge between life and death.
George Coşbuc - *The Poet*

In this Chapter we present some concepts regarding the exponential matrix, flows and related dynamical systems that have direct applications in science and engineering. Explicit links: (1) the connection between the exponential matrix and the solutions of linear systems of differential equations with constant coefficients; (2) the relationship between flows and vector fields; (3) the correlation between dynamical systems and vector fields. Suggestions for further reading: [1]-[9].

1.1 Exponential matrix

Let $\mathcal{M}_{n\times n}(\mathbb{R})$ be the set of square matrices of order n, endowed with a norm $\|\cdot\|$. Let I be the unit matrix of order n. For each matrix $A \in \mathcal{M}_{n\times n}(\mathbb{R})$, the power series

$$\sum_{k=0}^{\infty} \frac{A^k}{k!} = I + \frac{A}{1!} + \frac{A^2}{2!} + ...$$

is absolutely convergent and hence convergent. For a matrix of functions, A, the series is uniform convergent on each subset $\{A \mid ||A|| \leq a,\, a \in \mathbb{R}_+\}$. The sum of the previous series is called the *exponential matrix* and is denoted by e^A or $\exp A$.

If $\mathcal{O}$ means zero matrix, then $e^{\mathcal{O}} = I$. Also, the absolute convergence implies the fact that if $A, B \in \mathcal{M}_{n\times n}(\mathbb{R})$ are commutable matrices, i.e., $AB = BA$, then $e^A e^B = e^{A+B}$. It follows that the matrix e^A is invertible and its inverse is e^{-A}.

Here we are interested by the matrix function

$$\varphi : \mathbb{R} \to \mathcal{M}_{n\times n}(\mathbb{R}),\, \varphi(t) = e^{tA} = \sum_{k=0}^{\infty} \frac{t^k A^k}{k!},$$

also called the *exponential matrix*. The Cayley-Hamilton Theorem ensures that the exponential matrix e^{tA} is a polynomial of degree $n-1$, where n is the order of the matrix A. The coefficients of the polynomial in A are series of functions. Explicitly

$$e^{tA} = \sum_{k=0}^{\infty} \frac{t^k A^k}{k!} = \sum_{k=0}^{n-1} c_k(t) A^k.$$

Theorem 1.1.1. *The matrix function $\varphi(t) = e^{tA}$ is the solution of the Cauchy problem*

$$\frac{d\varphi}{dt}(t) = A\varphi(t),\, \varphi(0) = I.$$

Proof. The series

$$\varphi(t) = e^{tA} = \sum_{k=0}^{\infty} \frac{t^k A^k}{k!}, \text{ and } \sum_{k=0}^{\infty} \frac{d}{dt}\frac{t^k A^k}{k!} = A\sum_{k=0}^{\infty} \frac{t^k A^k}{k!}$$

converge absolutely and uniformly in any domain of $\mathbb{R} \times \mathcal{M}_{n\times n}(\mathbb{R})$ characterized by $|t| \leq b$ and $||A|| \leq a$. That is why there exists $\frac{d\varphi}{dt}(t)$ (derivative of the sum of initial series) and it is equal to the sum of the series formed deriving term by term. □

The matrices A and e^{tA} are commutable. This remark allows us to give a nice proof for

Theorem 1.1.2. *The matrix e^{tA} is invertible and its inverse is e^{-tA}.*

Proof. Let us consider the matrix $E(t) = e^{tA}e^{-tA}$, $t \in \mathbb{R}$. By derivation, we find

$$\frac{dE}{dt}(t) = Ae^{tA}e^{-tA} + e^{tA}(-A)e^{-tA} = \mathcal{O},\ \forall t \in \mathbb{R}.$$

It follows $E(t) = E(0) = I$. □

The direct calculation of the matrix e^{tA} is cumbersome or even impossible. That is why we explain some helpful techniques.

Case of a diagonal matrix For a diagonal matrix

$$D = \begin{pmatrix} \lambda_1 & 0 & \dots & 0 \\ 0 & \lambda_2 & \dots & 0 \\ \dots & \dots & \dots & \dots \\ 0 & 0 & \dots & \lambda_n \end{pmatrix},$$

we have

$$D^k = \begin{pmatrix} \lambda_1^k & 0 & \dots & 0 \\ 0 & \lambda_2^k & \dots & 0 \\ \dots & \dots & \dots & \dots \\ 0 & 0 & \dots & \lambda_n^k \end{pmatrix}$$

and hence

$$e^{tD} = \begin{pmatrix} e^{\lambda_1 t} & 0 & \dots & 0 \\ 0 & e^{\lambda_2 t} & \dots & 0 \\ \dots & \dots & \dots & \dots \\ 0 & 0 & \dots & e^{\lambda_n t} \end{pmatrix}.$$

Case of a Jordan cell Let J_p be a Jordan cell of order n_p, i.e.,

$$J_p = \begin{pmatrix} \lambda_p & 1 & 0 & \dots & 0 \\ 0 & \lambda_p & 1 & \dots & 0 \\ \dots & \dots & \dots & \dots & \dots \\ 0 & 0 & 0 & \dots & 1 \\ 0 & 0 & 0 & \dots & \lambda_p \end{pmatrix} = \lambda_p I_p + E_p,$$

where I_p is the unit matrix of order n_p, and the matrix

$$E_p = \begin{pmatrix} 0 & 1 & 0 & \dots & 0 \\ 0 & 0 & 1 & \dots & 0 \\ \dots & \dots & \dots & \dots & \dots \\ 0 & 0 & 0 & \dots & 1 \\ 0 & 0 & 0 & \dots & 0 \end{pmatrix}$$

is nilpotent of order n_p. Taking into account that the matrices I_p and E_p are commutable and computing the powers of E_p, it follows

$$e^{tJ_p} = e^{\lambda_p t I_p} e^{tE_p} = e^{\lambda_p t} \left[I_p + \frac{tE_p}{1!} + \frac{t^2 E_p^2}{2!} + ... + \frac{t^{n_p-1} E_p^{n_p-1}}{(n_p-1)!} \right]$$

$$= e^{\lambda_p t} \begin{pmatrix} 1 & \frac{t}{1!} & \frac{t^2}{2!} & ... & \frac{t^{n_p-1}}{(n_p-1)!} \\ 0 & 1 & \frac{t}{1!} & ... & \frac{t^{n_p-2}}{(n_p-2)!} \\ ... & ... & ... & ... & ... \\ 0 & 0 & 0 & ... & 1 \end{pmatrix}.$$

Generally, a Jordan matrix has the form

$$J = \begin{pmatrix} J_1 & 0 & ... & 0 \\ 0 & J_2 & ... & 0 \\ ... & ... & ... & ... \\ 0 & 0 & ... & J_k \end{pmatrix},$$

where J_p, $p = 1, ..., k$ are Jordan cells of orders n_p and $n_1 + ... + n_k = n$. It follows

$$e^{tJ} = \begin{pmatrix} e^{tJ_1} & 0 & ... & 0 \\ 0 & e^{tJ_2} & ... & 0 \\ ... & ... & ... & ... \\ 0 & 0 & ... & e^{tJ_k} \end{pmatrix}.$$

Case of diagonalizable and Jordanizable matrices First, we formulate a general result regarding the similar matrices. This gives rise to a consequence that is helpful to us in the calculations.

Theorem 1.1.3. *If S is a nonsingular matrix and if $S^{-1}AS = B$, then $e^{tA} = Se^{tB}S^{-1}$.*

Proof. The relation $A = SBS^{-1}$ implies $A^k = SB^kS^{-1}$. That is why

$$e^{tA} = \sum_{k=0}^{\infty} \frac{t^k A^k}{k!} = \sum_{k=0}^{\infty} \frac{t^k SB^kS^{-1}}{k!} = S\left(\sum_{k=0}^{\infty} \frac{t^k B^k}{k!}\right) S^{-1} = Se^{tB}S^{-1}.$$

□

Corollary 1.1.4. *(1) If A is a diagonalizable matrix, D is the diagonal matrix associated to A, and S is the diagonalizing matrix (the matrix of eigenvectors placed on columns), then $e^{tA} = Se^{tD}S^{-1}$.*

(2) If the matrix A has multiple eigenvalues and is reduced to the Jordan canonical form J, and S is the jordanizing matrix (the matrix of eigenvectors and principal vectors placed on columns), then $e^{tA} = Se^{tJ}S^{-1}$.

1.1.1 Connection with systems of linear differential equations

First, we refer to systems of linear differential equations with constant coefficients. Let $A \in \mathcal{M}_{n\times n}(\mathbb{R})$ be a real square matrix of order n.

Theorem 1.1.5. *The solution of the Cauchy problem $\dot{x}(t) = Ax(t)$, $x(0) = x_0$ is $x(t) = e^{tA}x_0$, $t \in \mathbb{R}$.*

Proof. The existence and uniqueness theorem together with the extension theorem show that the solution exists, is unique and defined on the whole set $\mathbb{R}$. The rest is just checks by calculations. □

Second, we refer to systems of linear differential equations with nonconstant coefficients.

Theorem 1.1.6. *The matriceal differential equation $\dot{X}(t) = A(t)X(t)$, $X(0) = X_0$ has the solution*

$$X(t) = \exp\left(\int_0^t A(\tau)d\tau\right) X_0$$

iff the matrices $A(t)$ and $\displaystyle\int_0^t A(\tau)d\tau$ commute.

Theorem 1.1.7. *The solution of Cauchy problem $\dot{X}(t) = A(t)X(t)$, $X(0) = X_0$ can be written in the form*

$$X(t) = \exp(\Omega(t))\, X_0\,,$$

where $\dot{\Omega}(t)$ and $A(t)$ are similar and $\Omega(0) = 0$. If the matrices $A(t)$ and $\Omega(t)$, or $\dot{\Omega}(t)$ and $\Omega(t)$, commute, then $\dot{\Omega}(t) = A(t)$.

Proof. Looking for a solution as $X(t) = \exp(\Omega(t))\, X_0$, we compute,

$$\dot{X}(t) = \left(\frac{d}{dt}\exp(\Omega(t))\right) X_0 = \int_0^1 \exp(\alpha\Omega(t))\,\dot{\Omega}(t)\,\exp((1-\alpha)\Omega(t))\; d\alpha\, X_0.$$

But $X_0 = \exp(-\Omega(t))\, X(t)$. Then,

$$\dot{X}(t) = \int_0^1 \exp(\alpha\Omega(t))\, \dot{\Omega}(t)\, \exp((-\alpha)\Omega(t))\; d\alpha\, X(t).$$

Comparing with the given system and applying the mean-value Theorem, we find a function $\alpha : [0, 1] \to \mathbb{R}$ such that

$$A(t) = \exp(\alpha(t)\Omega(t))\, \dot{\Omega}(t)\, \exp(-\alpha(t)\Omega(t))\,.$$

This means that the matrices $\dot{\Omega}(t)$ and $A(t)$ are similar.

If the matrices $A(t)$ and $\Omega(t)$ commute, then $A(t)$ and $\exp(\Omega(t))$ commute and there results $\dot{\Omega}(t) = A(t)$, as in the previous theorem. Similarly, if $\dot{\Omega}(t)$ and $\Omega(t)$ commute. □

Example 1.1.8. Find a fundamental matrix of the system of differential equations $\dot{x}(t) = x(t) + ty(t), \dot{y}(t) = tx(t) + y(t)$ making sure that the coefficient matrix

$$A(t) = \begin{pmatrix} 1 & t \\ t & 1 \end{pmatrix}, \; t > 0.$$

commutes with its integral.

The integral of the matrix $A(t)$ is found by elementwise integration. For simplicity, we take the lower boundary of integration to be zero. Then,

$$\int_0^t A(\tau)d\tau = \begin{pmatrix} t & \frac{t^2}{2} \\ \frac{t^2}{2} & t \end{pmatrix}$$

By computation, we verify $A(t) \cdot \int_0^t A(\tau)d\tau = \int_0^t A(\tau)d\tau \cdot A(t)$. So, the commutative property of the matrix product is true. Therefore, the fundamental matrix is given by

$$\Phi(t) = \exp\left(\int_0^t A(\tau)d\tau\right) = \exp\begin{pmatrix} t & \frac{t^2}{2} \\ \frac{t^2}{2} & t \end{pmatrix}.$$

We compute the matrix exponential by converting the matrix to the diagonal form. In this case, the eigenvalues and the eigenvectors depend on the variable t. Since the matrix of reduction to the diagonal form is given by

$$S = \begin{pmatrix} 1 & -1 \\ 1 & 1 \end{pmatrix},$$

we find

$$\Phi(t) = \exp \begin{pmatrix} t & \frac{t^2}{2} \\ \frac{t^2}{2} & t \end{pmatrix} = e^t \begin{pmatrix} \cosh \frac{t^2}{2} & \sinh \frac{t^2}{2} \\ \sinh \frac{t^2}{2} & \cosh \frac{t^2}{2} \end{pmatrix}.$$

Example 1.1.9. Consider the homogeneous linear differential system $\dot{x}(t) = A(t)x(t)$, where

$$A(t) = \begin{pmatrix} \frac{4}{t} & -\frac{4}{t^2} \\ 2 & -\frac{1}{t} \end{pmatrix}, \; t > 0.$$

Here the vectors $x^1(t) = {}^T[2t^2, t^3]$, $x^2(t) = {}^T[1, t]$ are particular solutions. Since

$$\begin{vmatrix} 2t^2 & 1 \\ t^3 & t \end{vmatrix} = t^3 > 0,$$

the solutions $x^1(t)$ and $x^2(t)$ are linearly independent. The general solution of the system is $x(t) = c_1 x^1(t) + c_2 x^2(t)$.

Let us write the solution using the exp matrix. Here, the idea of solving is based on knowing the solution

$$x(t) = \begin{pmatrix} 2t^2 & 1 \\ t^3 & t \end{pmatrix} \begin{pmatrix} c_1 \\ c_2 \end{pmatrix}.$$

Identifying, we obtain

$$\exp(\Omega(t)) = \begin{pmatrix} 2t^2 & 1 \\ t^3 & t \end{pmatrix}, \; t > 0.$$

1.2 Flows

Let D be an open subset in $\mathbb{R}^n$. A family of functions $\{T^t : D \to \mathbb{R}^n, t \in (-\epsilon, \epsilon)\}$ is called *local flow* on D or *local group with one parameter* of diffeomorphisms on D if the following conditions are satisfied:

(1) The function $T : (-\epsilon, \epsilon) \times D \to \mathbb{R}^n$, $(t, x) \to T^t(x)$ is of class C^∞.

(2) For each $t \in (-\epsilon, \epsilon)$, the function $T^t : D \to T^t(D)$ is a diffeomorphism.

(3) For each pair $t, s \in (-\epsilon, \epsilon)$, with $t + s \in (-\epsilon, \epsilon)$ and $T^s(x) \in D$, we have $T^{t+s}(x) = T^t(T^s(x))$, for each $x \in D$. It follows $T^0(x) = x$.

The adjective "local" refers to $t \in (-\epsilon, \epsilon)$. If $(-\epsilon, \epsilon) = \mathbb{R}$, i.e., $\epsilon = \infty$, then this adjective is replaced by "global".

A flow T^t defines a vector field $X(x) = \frac{d}{dt}T^t(x)|_{t=0}$ on D. Conversely, any vector field of class C^∞ defines a flow by its field lines. This thing will be proved in the following Section.

Let us consider the *affine group of transformations*

$$x^{i'} = a^i_j(t)x^j + a^i(t),\ \det(a^i_j(t)) \neq 0,\ t \in (-\epsilon, \epsilon),\ a^i_j(0) = \delta^i_j,\ a^i(0) = 0$$

on $\mathbb{R}^n$. The associated affine vector field is defined by

$$\frac{dx^{i'}}{dt}(0) = \frac{da^i_j}{dt}(0)x^j + \frac{da^i}{dt}(0).$$

Denoting $A = \left(\frac{da^i_j}{dt}(0)\right)$, $a = \left(\frac{da^i}{dt}(0)\right)$, we find $X(x) = Ax + a$. The flow generated by the vector field $X(x) = Ax + a$ consists in certain affine transformations.

1.3 Dynamical systems

The evolution over time of some physical systems can be modelled as dynamical systems.

Definition 1.3.1. *A semigrup G which acts on the space $\mathbb{R}^n$, in the sense that there exists a function $\Phi : G \times \mathbb{R}^n \to \mathbb{R}^n$ such that $\Phi_s \circ \Phi_t = \Phi_{sot}$ is called dynamical system. The function Φ is called flow.*

If G is a group, then we say that the dynamical system is *invertible.*

The discrete dynamical systems correspond to $G = \mathbb{N} \setminus \{0\} = \mathbb{N}_0$ or $G = \mathbb{Z}$ and they have the prototype a recursive sequence described by $\Phi : G \to G$, $\Phi^n = \Phi \circ \Phi^{n-1}$, $n \in \mathbb{N}_0$ or $x(t+1) = X(x(t)),\ t \in \mathbb{N}_0$.

The continuous dynamical systems correspond to $G = \mathbb{R}_+$ or $G = \mathbb{R}$ and they have the prototype described by a first order differential system of the form $\dot{x}(t) = X(x(t))$, where X is a vector field on $\mathbb{R}^n$ of class C^∞. The maximal curve that satisfies the Cauchy problem $\dot{x}(t) = X(x(t))$, $x(0) = x_0$ is defined on the interval $I_{x_0} = (T_-(x_0), T_+(x_0))$ and is called field line. Abandoning the index 0 and entering the set $W = \cup_{x \in \mathbb{R}^n} I_x \times \{x\} \subset \mathbb{R} \times \mathbb{R}^n$, we define the flow $\Phi : W \to \mathbb{R}^n$, $(t, x) \to \Phi(t, x)$, where $t \to \Phi(t, x)$ is an integral curve starting from the point x. In other words, a solution $\alpha = \alpha(t), \alpha(0) =$

x, $t \in (-\epsilon, \epsilon)$ of the differential system $\dot{x}(t) = X(x(t))$ generates a local diffeomorphism

$$T^t : U \subset \mathbb{R}^n \to \mathbb{R}^n,\ T^t(x) = \Phi(t,x),\ t \in I,$$

solution of the operatorial differential equation

$$\frac{dT^t}{dt} = X \circ T^t,\ T^0 = id,$$

called *local flow on* $\mathbb{R}^n$ generated be the vector field X (*local group with one parameter of diffeomorphisms*). The linear approximation $x' = x + tX(x)$ of $T^t(x)$ is called *infinitesimal transformation generated by the vector field* X.

The image set in $\mathbb{R}^n$ of the curve (field line, integral curve) $\alpha_x(t) = \Phi(t,x)$ is called *orbit* passing through the point $x \in \mathbb{R}^n$. A point $x \in \mathbb{R}^n$ which verifies $\Phi(t,x) = x$, for any $t \in I$, is called *equilibrium point* for the system. In this case the orbit passing through the point x is reduced to this point, $\{x\} = \alpha_x$.

If any solution $\alpha = \alpha(t), \alpha(0) = x$ is defined for each $x \in \mathbb{R}^n$ and each $t \in \mathbb{R}$, then the vector field X is called *complete*, and the flow is called *global* (*group with a parameter* of diffeomorphisms).

A graphical representation of the equilibrium points and of some generic orbits of the dynamical system is called the *phase portrait.*

To point things out, let's take the example $X(x) = x^3$, $\dot{x}(t) = x^3(t)$ defined on $\mathbb{R}$. Here we find the flow

$$\Phi(t,x) = \frac{x}{\sqrt{1 - 2x^2 t}},\ W = \{(t,x) \,|\, 2tx^2 < 1\},$$

$$T_-(x) = -\infty,\ T_+(x) = \frac{1}{2x^2}.$$

Example 1.3.1. We consider the linear vector field

$$X(x,y,z) = (x - y + z, 2y - z, z).$$

The field lines are solutions of the linear differential system (with constant coefficients)

$$\frac{dx}{dt} = x - y + z,\ \frac{dy}{dt} = 2y - z,\ \frac{dz}{dt} = z.$$

The third equation has the general solution $z = ce^t$, $t \in \mathbb{R}$. Replacing in the second equation, we find a linear differential equation $\frac{dy}{dt} = 2y-$

ce^t, with the general solution $y = be^{2t} + ce^t$, $t \in \mathbb{R}$. Introducing y and z in the first equation, we get $\frac{dx}{dt} = x - be^{2t}$, with the general solution $x = ae^t - be^{2t}$, $t \in \mathbb{R}$. Obviously $(0,0,0)$ is the only equilibrium point.

A modified flow generated by the previous vector field is

$$T^t : \mathbb{R}^3 \to \mathbb{R}^3,\ T^t(x,y,z) = (xe^t - ye^{2t}, ye^{2t} + ze^t, ze^t),\ t \in \mathbb{R}.$$

In order to study the effect of the flow on the volume, we use the evolution velocity, i.e., the divergence of the vector field X,

$$\operatorname{div}(X) = \frac{\partial X^1}{\partial x} + \frac{\partial X^2}{\partial y} + \frac{\partial X^3}{\partial z} = 1 + 2 + 1 = 4.$$

Since $\operatorname{div}(X) > 0$, the flow associated to this vector field increases the volume.

Example 1.3.2. Study the completeness of the non-linear vector field

$$X(x,y,z) = (xz, yz, -(x^2 + y^2)).$$

The orbits of the vector field X, solutions of the symmetric differential system

$$\frac{dx}{xz} = \frac{dy}{yz} = \frac{dz}{-(x^2+y^2)},$$

are $x = c_1 y, x^2 + y^2 + z^2 = c_2$. These orbits can thus be parameterized

$$x = \sqrt{c_2}\cos u \cos v,\ y = \sqrt{c_2}\cos u \sin v,\ z = \sqrt{c_2}\sin u,\ u \in \mathbb{R},$$

$\cos v = c_1 \sin v$. We parametrize again $u = u(t)$ such that $\frac{dz}{dt} = -(x^2 + y^2)$. Since $\frac{dz}{dt} = \frac{dz}{du}\frac{du}{dt}$, it follows $\sqrt{c_2}\cos u \frac{du}{dt} = -c_2 \cos^2 u$ and hence $\frac{du}{\cos u} = -\sqrt{c_2}\,dt$. So,

$$\ln\left|\operatorname{tg}\left(\frac{\pi}{4} - \frac{u}{2}\right)\right| = -\sqrt{c_2}\,t$$

or

$$\operatorname{tg}\left(\frac{\pi}{4} - \frac{u}{2}\right) = e^{-\sqrt{c_2}\,t},\ t \in \mathbb{R}.$$

Consequently, the vector field X is complete.

Shortcut The orbits of X belong to the spheres $x^2 + y^2 + z^2 = c_2$, and these surfaces are compact. Therefore the vector field X is complete.

1.4 The almost linear expression of a vector field

We consider a real function f of class C^1 defined on a convex neighbourhood V of a point $x_0 \in \mathbb{R}^n$. Then it appears the almost linear expression

$$f(x) = f(x_0) + \int_0^1 \frac{d}{dt} f(x_0 + t(x - x_0))dt$$

$$= f(x_0) + \int_0^1 f_{x^i}(x_0 + t(x - x_0))(x^i - x_0^i)dt = f(x_0) + g_i(x)(x^i - x_0^i),$$

where $g_i(x_0) = f_{x^i}(x_0)$. This result is known as Hadamard's Lemma.

Extending this remark to a vector field X of class C^1, with the property $X(x_0) = 0$, there is a quadratic matrix $A(x)$ such that $X(x) = A(x)x$, $x \in V$. We combine with the remark that a Cauchy problem $\dot{x}(t) = A(x(t))x(t)$, $x(t_1) = x_1$ is equivalent to an integral equation $x(t) = x(t_1) + \int_{t_1}^t A(x(s))x(s)ds$.

1.5 Self-evaluation problems

Problem 1. *Show that $T(t, x) = e^t(1 + x) - 1$ is a flow and write the dynamical system that corresponds to this flow.*

Solution. We show that $T(0, x) = x$ and $T(t + s, x) = T(t, T(s, x))$ (or, if we denote $T(t, x)$ with $T^t(x)$, then $T^0(x) = id(x) = x$ and $T^{t+s}(x) = (T^t \circ T^s)(x)$). The relationships are obviously true, so the application T is a flow.

Since

$$\frac{d}{dt}T^t(x)|_{t=0} = \frac{d}{dt}\alpha_x(t)|_{t=0} = \dot{\alpha}_x(0) = X(\alpha_x(0)) = X(x),$$

the vector field corresponding to this flow is

$$X(x) = \frac{d}{dt}T^t(x)|_{t=0} = \frac{d}{dt}(e^t(1 + x) - 1)|_{t=0} = 1 + x,$$

and the associated differential equation is $\dot{x}(t) = x(t) + 1$.

Problem 2. *Describe the flow generated by the vector field $X(x) = x^3$ on $\mathbb{R}$.*

Solution. X being a vector field on $\mathbb{R}$, the field lines are described by those functions $\alpha : I \to \mathbb{R}$, $\alpha(t) = x(t)$ which verify the ODE $\dot{x}(t) = X(x(t))$, i.e., $\dot{x}(t) = x^3(t)$. By integration, we find the solution $x^2 = \dfrac{1}{c-2t}$, $c \in \mathbb{R}$. The field lines are the functions

$$\alpha(t) = x(t) = \frac{1}{\pm\sqrt{c-2t}}, \quad c \in \mathbb{R}.$$

Let $x \in \mathbb{R}_+$. We are looking for that field line that "leaves" from x, i.e., $\alpha(0) = x$, hence $c = \dfrac{1}{x^2}$. Therefore, the flow associated with the vector field in the statement is the application $T^t(x) = \dfrac{x}{\sqrt{1-2tx^2}}$.

Problem 3. *Find out the flow generated by the linear vector field*

$$X(x, y, z) = (x + y - z, x - y + z, -x + y + z)$$

and investigate its effect on volume.

Solution. The field lines are the solutions of the differential system

$$\dot{\alpha}(t) = X((\alpha(t)), \quad \begin{cases} \dot{x} = x + y - z \\ \dot{y} = x - y + z \\ \dot{z} = -x + y + z. \end{cases}$$

This is a linear system, that can be solved by eigenvalues and eigenvectors method.

The eigenvalues and eigenvectors of the associated matrix are: $\lambda_1 = 1$, whose eigenvector is $v_1 = (1, 1, 1)$; $\lambda_2 = 2$, with $v_2 = (1, 0, -1)$ and $\lambda_3 = -2$, with $v_3 = (1, -2, 1)$. Then the general solution of the ODE system is

$$\begin{pmatrix} x \\ y \\ z \end{pmatrix} = c_1 e^t \begin{pmatrix} 1 \\ 1 \\ 1 \end{pmatrix} + c_2 e^{2t} \begin{pmatrix} 1 \\ 0 \\ -1 \end{pmatrix} + c_3 e^{-2t} \begin{pmatrix} 1 \\ -2 \\ 1 \end{pmatrix}.$$

Equivalently, the field lines are

$$\alpha(t) = (c_1 e^t + c_2 e^{2t} + c_3 e^{-2t}, c_1 e^t - 2c_3 e^{-2t}, c_1 e^t - c_2 e^{2t} + c_3 e^{-2t}),$$

$c_1, c_2, c_3 \in \mathbb{R}$. The only equilibrium point is $(0, 0, 0)$.

By definition, the flow is the map T^t which associates to each arbitrary point (x, y, z) of $\mathbb{R}^3$ the field line $\alpha_{(x,y,z)}(t)$, satisfying $\alpha_{(x,y,z)}(0) = (x, y, z)$. For $t = 0$, we obtain the linear algebraic system

$$c_1 + c_2 + c_3 = x,\ c_1 - 2c_3 = y,\ c_1 - c_2 + c_3 = z,$$

with unique solution $c_1 = \dfrac{x+y+z}{3}$, $c_2 = \dfrac{x-z}{2}$, $c_3 = \dfrac{x-2y+z}{6}$. Hence the flow is

$$T^t(x, y, z) = \left(\frac{x+y+z}{3} e^t + \frac{x-z}{2} e^{2t} + \frac{x-2y+z}{6} e^{-2t}, \right.$$

$$\frac{x+y+z}{3} e^t - 2\frac{x-2y+z}{6} e^{-2t}, \frac{x+y+z}{3} e^t - \frac{x-z}{2} e^{2t}$$

$$\left. + \frac{x-2y+z}{6} e^{-2t} \right).$$

To determine the effect of the flow on the volume, we calculate the divergence of the vector field X, i.e.,

$$\mathrm{div}(X) = \frac{\partial X^1}{\partial x} + \frac{\partial X^2}{\partial y} + \frac{\partial X^3}{\partial z} = 1 - 1 + 1 = 1.$$

Since $\mathrm{div}(X) > 0$, the flow increases the volume.

Problem 4. *Study the effect of flow generated by the non-linear vector field*

$$X(x, y, z) = (x^2 + y^2, 2xy, -z^2)$$

on the volume. Write the straightening diffeomorphism and the corresponding infinitesimal transformation.

Solution. The field lines are the solutions of differential system

$$\dot{x} = x^2 + y^2,\ \dot{y} = 2xy,\ \dot{z} = -z^2.$$

The dynamical system has only one equilibrium point, $(0, 0, 0)$.

We write the associated symmetric differential system

$$\frac{dx}{x^2 + y^2} = \frac{dy}{2xy} = \frac{dz}{-z^2},$$

to find the first integrals. The first two quotients defines an homogeneous differential equation, with unknown $y = y(x)$. With the change of unknown function $u = \frac{y}{x}$, we obtain a separable variables equation $\frac{1+u^2}{u-u^3}du = \frac{dx}{x}$, with the unknown $u = u(x)$. The fraction in the left member is broken down into simple fractions. As a result of integration, we find the equality $\frac{u}{1-u^2} = xc$, $c \in \mathbb{R}$, whence it follows $\frac{y}{x^2-y^2} = c_1$, $c_1 \in \mathbb{R}$. The expression $\frac{y}{x^2-y^2}$ defines a first integral.

To find another first integral, we use the method of integrable combinations. By adding the numerators and denominators of the first two reports, we reach $\frac{d(x+y)}{(x+y)^2} = \frac{dz}{-z^2}$. Denote $x+y = v$. Integrating, we obtain $\frac{1}{v} + \frac{1}{z} = c$, $c \in \mathbb{R}$, i.e., $\frac{1}{x+y} + \frac{1}{z} = c_2$, $c_2 \in \mathbb{R}$. In this way, the expression $\frac{1}{x+y} + \frac{1}{z}$ defines also a first integral.

We study the functional independence of the two first integrals. The Jacobi matrix is

$$\begin{pmatrix} -\frac{2xy}{(x^2-y^2)^2} & \frac{x^2+y^2}{(x^2-y^2)^2} & 0 \\ -\frac{1}{(x+y)^2} & -\frac{1}{(x+y)^2} & -\frac{1}{z^2} \end{pmatrix}.$$

The rank of the matrix is two on $\mathbb{R}^3 \setminus Oz$. Hence, on this set, the two first integrals are functional independent.

The implicit Cartesian equations of field lines are

$$\begin{cases} \frac{y}{x^2-y^2} = c_1 \\ \frac{1}{x+y} + \frac{1}{z} = c_2, \quad c_1, c_2 \in \mathbb{R}. \end{cases}$$

The straightening diffeomorphism (change of variables) is

$$\phi : \mathbb{R}^3 \to \mathbb{R}^3,\ \phi(x,y,z) = (\tilde{x}, \tilde{y}, \tilde{z}), \begin{cases} \dot{\tilde{x}} = 0 \\ \dot{\tilde{y}} = 0 \\ \dot{\tilde{z}} = 1. \end{cases}$$

We define $\tilde{x} = \frac{y}{x^2-y^2}$ and $\tilde{y} = \frac{1}{x+y} + \frac{1}{z}$ (the right hand members are constant along the field lines, so their derivatives with respect to the parameter t is equal to zero).

We are looking for $\tilde{z}$. From $\dfrac{d\tilde{z}}{dt} = 1$, it follows $\dfrac{d\tilde{z}}{dz} \cdot \dfrac{dz}{dt} = 1$, i.e., $\dfrac{d\tilde{z}}{dz} \cdot (-z^2) = 1$, equation with separable variables which by integration leads to $\tilde{z} = \dfrac{1}{z}$.

Let $A = \{(x, y, z) \,|\, x - y = 0 \text{ or } x + y = 0 \text{ or } z = 0\}$. It follows the straightening diffeomorphism

$$\phi : \mathbb{R}^3 \setminus A \to \mathbb{R}^3,\ \phi(x, y, z) = \left(\frac{y}{x^2 - y^2},\ \frac{1}{x + y} + \frac{1}{z},\ \frac{1}{z} \right).$$

Remark. By a diffeomorphism $x^{i'} = x^{i'}(x^i)$, $i = 1, 2, 3$, the components X^i of a vector field are changed after the rule (tensorial law)

$$X^{i'} = \frac{\partial x^{i'}}{\partial x^i} X^i.$$

The infinitesimal transformation determined by the vector field X is

$$(x', y', z') = (x, y, z) + tX(x, y, z),$$

i.e.,

$$x' = x + t(x^2 + y^2),\ y' = y + 2txy,\ z' = z - tz^2, \quad t \in (-\varepsilon, \varepsilon).$$

Problem 5. *Find the straightening diffeomorphism and the infinitesimal transformation for the vector field* $X(x, y, z) = (z, -xy, 2xz)$.

Solution. The field lines are the solutions of the differential system

$$\dot{x} = z,\ \dot{y} = -xy,\ \dot{z} = 2xz.$$

The equilibrium points are the solutions of the algebraic system

$$z = 0,\ xy = 0,\ 2xz = 0,$$

i.e., all the points of the form $(\alpha, 0, 0)$ and $(0, \alpha, 0)$, with $\alpha \in \mathbb{R}$. In other words, the axes Ox and Oy consist of equilibrium points.

To find the first integrals, we use the associated symmetric system

$$\frac{dx}{z} = \frac{dy}{-xy} = \frac{dz}{2xz}.$$

The first ratio and the last one determine a differential equation with separable variables, which, by integration, leads to $x^2 - z = c_1, \quad c_1 \in \mathbb{R}$.

The second and third ratios also determine an equation with separable variables, with the implicit solution $zy^2 = c_2, \quad c_2 \in \mathbb{R}$.

The functional independence of the two first integrals, $x^2 - z$ and zy^2, is established using the rank of the Jacobi matrix

$$\begin{pmatrix} 2xy & 0 & -1 \\ 0 & 2yz & y^2 \end{pmatrix}.$$

This matrix has rank 1 on the axis Oy and for the points in the plane xOz, and in the other points its rank is 2, so the two first integrals found are functionally independent on $\mathbb{R}^3 \setminus Oy \cup xOz$. Cartesian equations of orbits are $x^2 - z = c_1, zy^2 = c_2$.

The straightening diffeomorphism is the map $\phi : \mathbb{R}^3 \to \mathbb{R}^3$, $\phi(x, y, z) = (\tilde{x}, \tilde{y}, \tilde{z})$, where $\dot{\tilde{x}} = 0, \dot{\tilde{y}} = 0, \dot{\tilde{z}} = 1$.

Define $\tilde{x} = x^2 - z$ and $\tilde{y} = zy^2$, using the first integrals, which are constant along the field lines, so their derivative with respect to the parameter t equal to zero.

Let us find $\tilde{z}$. From the condition $\dfrac{d\tilde{z}}{dt} = 1$, it follows $\dfrac{d\tilde{z}}{dz} \cdot \dfrac{dz}{dt} = 1$, i.e., $\dfrac{d\tilde{z}}{dz} \cdot 2xz = 1$. We express the variable x in function of z, from the Cartesian equations of field lines (using the first integrals). It follows $x = \sqrt{z + c_1}$, with the condition $z + c_1 > 0$. The equation for finding $\tilde{z}$ becomes $d\tilde{z} = \dfrac{1}{2z\sqrt{z + c_1}}\, dz$, which has separable variables. For the calculation of the integral of the right member, a change of variable is made, $\dfrac{1}{\sqrt{z + c_1}} = u$, which ultimately leads to the result

$$\tilde{z} = \frac{1}{2\sqrt{x^2 - z}} \ln \frac{|x| - \sqrt{x^2 - z}}{|x| + \sqrt{x^2 - z}}.$$

The straightening diffeomorphism is the map

$$\phi : D \subset \mathbb{R}^3 \to \mathbb{R}^3,\ \phi(x, y, z) = \left(x^2 - z, zy^2, \frac{1}{2\sqrt{x^2 - z}} \ln \frac{|x| - \sqrt{x^2 - z}}{|x| + \sqrt{x^2 - z}} \right).$$

The infinitesimal transformation determined by X is $(x', y', z') = (x, y, z) + tX(x, y, z)$, i.e., $x' = x + tz, y' = y - txy, z' = z + 2txz, \quad t \in (-\varepsilon, \varepsilon)$.

Problem 6. *Study the completeness of the vector field*

$$X(x, y, z) = (xz, yz, -x^2 - y^2).$$

Solution. If the differential system associated with the vector field admits a global first integral whose constant-level surfaces are compact, then the vector field is complete, that is the definition domain of field lines is the whole $\mathbb{R}$.

The first integrals can be found using the symmetric system

$$\frac{dx}{xz} = \frac{dy}{yz} = \frac{dz}{-x^2 - y^2}.$$

The first two ratios give a differential equation with separable variables. The solution $\dfrac{x}{y} = c_1, \quad x \neq 0, \ y \neq 0$, represents a family of portions of planes.

For the second first integral, we use the method of integrable combinations. We amplify the ratios by x, y, and z, respectively, then add the nominators and denominators respectively. We obtain

$$\frac{xdx}{x^2 z} = \frac{ydy}{y^2 z} = \frac{zdz}{-zx^2 - zy^2} = \frac{xdx + ydy + zdz}{0},$$

and hence $x^2 + y^2 + z^2 = c_2$, which represents a family of spheres with the center in origin. Therefore, the orbits are circles or equilibrium points.

Since the orbits are contained in the spheres $x^2 + y^2 + z^2 = c_2$, which are compact surfaces, it turns out that X is a complete vector field.

Problem 7. *We consider the Lorenz vector field*

$$X(x) = (x^2 - x^1, rx^1 - x^2 - x^1x^3, x^1x^2 - x^3), \quad r > 0.$$

Find the linear transformations to which it is invariant.

Solution. Let

$$x^{1'} = a_1^{1'} x^1 + a_2^{1'} x^2 + a_3^{1'} x^3,$$

$$x^{2'} = a_1^{2'} x^1 + a_2^{2'} x^2 + a_3^{2'} x^3,$$

$$x^{3'} = a_1^{3'} x^1 + a_2^{3'} x^2 + a_3^{3'} x^3,$$

where $a_i^{i'} \in \mathbb{R}$, $\forall i', i \in \{1,2,3\}$, be the equations of a linear transformation (change of coordinates). When changing the coordinates, the components of a vector field change according to the law

$$X^{i'} = \frac{\partial x^{i'}}{\partial x^i} X^i, \ \forall i', i \in \{1,2,3\}.$$

The invariance assumes that the expressions of the field components with respect to the new coordinates coincide with those with the initial coordinates, that is

$$\begin{cases} X^{1'}(x^{1'}, x^{2'}, x^{3'}) = x^{2'} - x^{1'} \\ X^{2'}(x^{1'}, x^{2'}, x^{3'}) = rx^{1'} - x^{2'} - x^{1'}x^{3'} \\ X^{3'}(x^{1'}, x^{2'}, x^{3'}) = x^{1'}x^{2'} - x^{3'}. \end{cases}$$

First relation (law of changes of components) becomes

$$x^{2'} - x^{1'} = \frac{\partial x^{1'}}{\partial x^1} X^1 + \frac{\partial x^{1'}}{\partial x^2} X^2 + \frac{\partial x^{1'}}{\partial x^3} X^3,$$

whence

$$a_1^{2'} x^1 + a_2^{2'} x^2 + a_3^{2'} x^3 - a_1^{1'} x^1 - a_2^{1'} x^2 - a_3^{1'} x^3 = a_1^{1'}(x^2 - x^1)$$

$$+ a_2^{1'}(rx^1 - x^2 - x^1x^3) + a_3^{1'}(x^1x^2 - x^3).$$

Identifying with respect to x^1, x^2, x^3, we find $a_2^{1'} = a_1^{2'} = a_3^{2'} = a_3^{1'} = 0$ and $a_1^{1'} = a_2^{2'}$. The equations of the linear transformation become

$$x^{1'} = a_1^{1'} x^1, \ x^{2'} = a_1^{1'} x^2, \ x^{3'} = a_1^{3'} x^1 + a_2^{3'} x^2 + a_3^{3'} x^3.$$

Repeating the procedure for the other two relationships, all coefficients $a_i^{i'} \in \mathbb{R}$, $\forall i', i \in \{1,2,3\}$ are determined. The linear transformations are the identity $x^{1'} = x^1, x^{2'} = x^2, x^{3'} = x^3$, and the symmetry $x^{1'} = -x^1, x^{2'} = -x^2, x^{3'} = x^3$ with respect to Ox^3.

Bibliography

[1] V. I. Arnold, *Geometrical Methods in the Theory of Ordinary Differential Equations*, Springer-Verlag, New York, 1983.

[2] E. Hairer, C. Lubich, G. Wanner, *Geometric Numerical Integration*, Springer, 2002.

[3] C.-G. Lefter, *Differential Equations and Dynamical Systems* (in Romanian), Alexandru Myller Editorial House, Iaşi, 2006.

[4] C. Udriste, *Geometric Dynamics*, Mathematics and Its Applications, 513, Kluwer Academic Publishers, Dordrecht, Boston, London, 2000.

[5] C. Udriste, *Differential Geometry, Differential Equations* (in Romanian), Universitext 15, Geometry Balkan Press, Bucharest, 2002.

[6] C. Udriste, L. Matei, *Lagrange-Hamilton Theories* (in Romanian), Monographs and Textbooks 8, Geometry Balkan Press, Bucharest, 2008.

[7] I. Vrabie, *Differential Equations* (in Romanian), MATRIXROM Publishing House, Bucharest, 2000.

[8] R. M. Wilcox, *Exponential operators and parameter differentiation in Quantum Physics*, Journal of Mathematical Physics. 8 (4) (1967); https://doi.org/10.1063/1.1705306

[9] E. T. Whittaker, *A Treatise on The Analytical Dynamics of Particles & Rigid Bodies*, Cambridge University Press, 1989.

Chapter 2

INVARIANT SETS, CONSERVATION AND DISSIPATION

Motto:
What I have told you is enow!
You swore on shields your oath of love
For Dacia! Might resides in you
That they, the gods, are far above,
Our foes - at a stone's throw!

George Coşbuc - *The Poet*

In this Chapter we present some concepts (dynamical systems, flows, Liouville theorem, invariant sets, preserved quantities, dissipated quantities, derivation along a trajectory, time discretization, Poincaré map, phytoplankton dynamical system) related to dynamical systems with direct applications in science, engineering and economics. We consider these topics to be essential for understanding the concept of dynamical system. Suggestions for further reading: [1]-[8].

2.1 Liouville Theorem

The time evolution of some physical systems can be modelled as dynamical systems. The continuous dynamical systems are modelled by first order differential systems of the form $\dot{x}(t) = X(x(t))$, where X is a vector field on $\mathbb{R}^n$ of class C^∞.

Remark 2.1.1. *Suppose that $X(x), x \in \mathbb{R}^n$ is a local Lipschitzian vector field. In case that the differential system $\dot{x}(t) = X(x(t))$ does not have global solutions (defined on the whole $\mathbb{R}$), and we pursue properties that do not depend on the parametrization of orbits, then we can move on topologically equivalent system*

$$\dot{x}(t) = \frac{1}{1 + ||X(x(t))||} X(x(t)).$$

The orbits of this system coincide geometrically with the orbits of the initial system and the solutions are defined globally because the right member of the new system is bounded. Moving to the new dynamic system corresponds to a change of the parametrization of the orbits of the initial differential system.

As example, we analyse the ODE $\dot{x}(t) = 1 + x^2(t)$. The general solution is $x(t) = \tan(t + c)$ and is not defined on the whole $\mathbb{R}$. More precisely, $t + c \in I_k = (-\frac{\pi}{2} + k\pi, \frac{\pi}{2} + k\pi), k \in \mathbb{Z}$. The global orbit is $im(x(t)) = (-\infty, \infty)$. The normed equation, in the previous sense, is $\dot{x}(t) = \frac{1+x^2(t)}{2+x^2(t)}$. Now the general solution is implicitly defined by $x(t) + \arctan(x(t)) = t + c$, and is defined on the whole $\mathbb{R}$. These graphs are obtained from the previous graphs by a rotation of angle $\frac{-\pi}{4}$. The global orbit is the same $im(x(t)) = (-\infty, \infty)$.

Theorem 2.1.1 (Liouville Theorem). *Let $X(x)$ be a vector field of class C^1 on $\mathbb{R}^n$ which determines the dynamical system $\dot{x}(t) = X(x(t))$, and $x' = \Phi(t, x)$ the associated flow. Let Ω be a measurable and bounded subset of $\mathbb{R}^n$ and $V = \int_\Omega dx$, the volume of the domain Ω, where $dx = dx^1...dx^n$ is the volume element. If $\Omega(t) = \Phi(t, \Omega)$ and $V(t) = \int_{\Omega(t)} dx'$, then*

$$\dot{V}(t) = \int_{\Omega(t)} div\, X(x) dx.$$

Proof. The change of variables in the multiple integral leads to

$$V(t) = \int_{\Omega(t)} dx' = \int_\Omega \det(d\Phi_t(x)) dx.$$

Since $d\Phi_t = I + tdX + \mathcal{R}(t)$, where $\mathcal{R}(t)$ contains the terms of at least degree two in the variable t, we deduce that

$$V(t) = \int_\Omega (1 + tr(dX)t + \mathcal{P}(t))dx.$$

Taking the derivative with respect to t and setting $t = 0$, we find

$$\dot{V}(0) = \int_\Omega tr(dX)dx.$$

We replace Ω with $\Omega(t)$ and we repeat the reasoning. We find that the previous relation takes place for any t and not only for $t = 0$. □

Corollary 2.1.2. *Let X be a vector field of class C^1. (1) If X is solenoidal, i.e., $div\, X = 0$, then the flow generated by X preserves the volume. (2) If $div\, X > 0$, then the flow increases the volume (it is expansive). (3) If $div\, X < 0$, then the current decreases the volume (it is contractive).*

Hence, $\operatorname{div} X$ represents the evolution velocity of volume in a flow.

2.2 Invariant sets

Any C^∞ vector field X on $\mathbb{R}^n$ determines an autonomous differential system (dynamical system) $\dot{x}(t) = X(x(t))$. A subset $S \subset \mathbb{R}^n$ is called *invariant* with respect to a dynamical system if for each trajectory $x(t)$ we have

$$x(t) \in S \Rightarrow x(\tau) \in S, \ \forall \tau \geq t.$$

This definition is equivalent to: (1) if a trajectory enters in the set S or starts from a point of the set S, it rests in S; (2) trajectories can enter S through the border ∂S, but can never exit outside S.

Example 2.2.1. (1) If x_0 is an equilibrium point, i.e., $X(x_0) = 0$, then $S = \{x_0\}$ is an invariant set. (2) Any trajectory or a union of trajectories is an invariant set. For example,

$$S = \{x(t) |\, x(0) \in D,\, t \geq 0,\, \dot{x}(t) = X(x(t))\}$$

is an invariant set. (3) Let $\dot{x}(t) = Ax(t)$ be a differential linear system with constant coefficients: (i) if $Av_i = \lambda_i v_i$, then $S = span\{v_1, ..., v_k\}$ is an invariant set; (ii) if $w^T A = \lambda w^T$, $\lambda \leq 0$, then $S = \{z \,|\, 0 \geq w^T z \geq a\}$ is an invariant set.

2.2.1 Non-negative orthant invariance

An usual question in the applications is: which linear dynamical systems with constant coefficients preserve the first orthant?

Theorem 2.2.2. *Let $A = (a_{ij})$ be a square matrix. The non-negative orthant $\mathbb{R}^n_+$ is invariant with respect to the dynamical system $\dot{x}(t) = Ax(t)$ if and only if $a_{ij} \geq 0$ for $i \neq j$.*

Proof. We use the identity

$$x(t) = e^{tA}\, x(0) = \lim_{n\to\infty} \left(I + \frac{t}{n} A\right)^n x(0).$$

Suppose that $a_{ij} \geq 0$ for $i \neq j$. Since for n enough large, the matrix $I + \frac{t}{n}A$ has all its elements non-negative, it follows that the matrix $\left(I + \frac{t}{n}A\right)^n$ has all elements non-negative and hence the limit $x(t)$ has all its elements non-negative.

Suppose there exists a pair of indices $i \neq j$ such that $a_{ij} < 0$. We fix $x(0) = e_j$. For $h > 0$, small, we have the approximation $x(h) \cong e_j + hAe_j$. Particularly $x(h)_i \cong ha_{ij} < 0$, i.e., $x(h) \notin \mathbb{R}^n_+$. $\square$

2.3 Conservation laws

We consider an autonomous dynamical system $\dot{x}(t) = X(x(t))$, where X is a vector field at least of class C^1. Let $F : \mathbb{R}^n \to \mathbb{R}$ be a nonconstant function.

If for each trajectory $x(t)$ the composition $F(x(t))$ is constant, then the function F is called *integral of motion* or *first integral* or *conserved quantity* (with respect to the dynamical system).

Obviously, the same constant can correspond to several trajectories. If the function F is of class C^1, then F is a first integral if and only if

$$\frac{d}{dt}F(x(t)) = \langle \nabla F(x(t)), X(x(t))\rangle = 0,$$

for any trajectory $x(t)$. The expression $\frac{d}{dt}F(x(t))$ is called the derivative of the function F along a trajectory $x(t)$.

The best known examples are: (1) the total energy of a conservative mechanical system (i.e., without loss of energy); (2) the total angular momentum around an axis of an isolated system; (3) the total fluid in a closed system.

Let $F : \mathbb{R}^n \to \mathbb{R}$ be a C^1 regular first integral of the dynamical system $\dot{x}(t) = X(x(t))$. The constant level sets (hypersurfaces) M_c : $F(x) = c$ are invariant sets because the trajectories are included in such subsets. Indeed, the vector ∇F is normal to each constant level hypersurface and hence, the vector field X is tangent to each constant level hypersurface.

If $\frac{d}{dt}F(x(t)) > 0$, then the composed function $F(x(t))$ is strictly increasing. If $\frac{d}{dt}F(x(t)) < 0$, then the composed function $F(x(t))$ is strictly decreasing. If $\frac{d}{dt}F(x(t)) = 0, \forall x(t)$, then the function $F(x)$ is a first integral; if F is regular, the vector field $X(x)$ is tangent to each hypersurface $M_c : F(x) = c$ or the vector fields ∇F and X are orthogonal.

The trajectories of a conservative mechanical system are in the constant level hypersurfaces of the energy.

Example 2.3.1. A nonlinear conservative mechanical system is presented in Fig.1. The Newton equation (associated differential equation) is

$$m\ddot{q}(t) = -X(q(t)),$$

where $m > 0$ is the mass, $q(t)$ is the displacement and $X(q)$ is the returning force, that is, the characteristic of the spring with $X(0) = 0$. Denoting $x = (x^1, x^2) = (q, \dot{q})$, we can write

$$\dot{x}(t) = (\dot{q}(t), \ddot{q}(t)) = \left(x^2(t), -\frac{1}{m}\, X(x^1(t))\right).$$

The potential energy stored in the spring is $U(q) = \displaystyle\int_0^q X(u)du$, and the total energy (kinetic energy + potential energy) has the expression

$$E(x) = \frac{m}{2}\dot{q}^2 + U(q).$$

The total energy is a first integral of motion. Indeed,

$$\frac{d}{dt}E(x(t)) = m\dot{q}\ddot{q} + X(q)\dot{q} = 0.$$

Theorem 2.3.2. *Let $\dot{Y} = A(Y)\, Y$ be a dynamical system, where the unknown Y can be a vector or a matrix (not necessarily quadratic). If the matrix $A(Y)$ is antisymmetric, then the quadratic form $I(Y) =$*

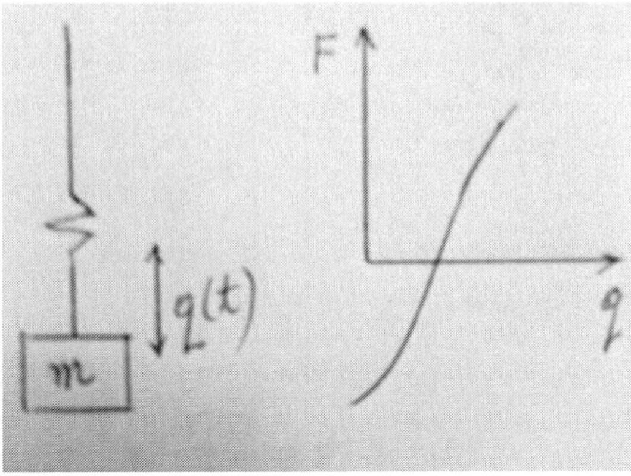

Figure 2.1: Nonlinear mechanical system

Y^TY is invariant (first integral). Particularly, if the initial value $Y(0) = Y_0$ consists in orthonormal columns, i.e., $Y_0^T Y_0 = I$, then the columns of the matrix solution $Y(t)$ are orthonormal.

Proof. We compute the derivative of the function $I(Y)$ on the direction H. We find $I'(Y)H = Y^T H + H^T Y$. We replace $H = A(Y)Y$. We deduce $I'(Y)A(Y)Y = Y^T A(Y)Y + (A(Y)Y)^T Y = 0$. □

Theorem 2.3.3. *Let be given the Cauchy problem $\dot{Y} = A(Y)\,Y$, $Y(0) = Y_0$, where $A(Y)$ and the unknown Y are quadratic matrices. If the matrix $A(Y)$ has vanishing trace, then the determinant $F(Y) = \det Y$ is a first integral (invariant).*

Proof. We have

$$\det(Y + \varepsilon AY) = \det(I + \varepsilon A)\det Y = (1 + \varepsilon trA + \mathcal{O}(\varepsilon^2))\det Y.$$

It follows $F'(Y)AY = trA\,\det Y$. Then $F'(Y)AY = 0$. □

2.3.1 Invariance criterion

We consider a C^1 vector field X and a C^1 real function F on $\mathbb{R}^n$.

Theorem 2.3.4. *We associate the dynamical system $\dot{x}(t) = X(x(t))$. If $F(x) = 0$ implies $\nabla F(x(t))^T X(x) < 0$, then the sublevel set $C = \{x\,|\,F(x) \leq 0\}$ is invariant with respect to the dynamical system.*

Proof. All trajectories, which reach the boundary ∂C of the set C, or starting from the boundary enters the set C. To show this, suppose that there is a trajectory $x(t)$ that satisfies $x(t) \in C$, $x(s) \notin C$, $t \leq s$. Then the differentiable function $g(\tau) = F(x(\tau))$ satisfies $g(t) \leq 0$, $g(s) > 0$. On the other hand, any such a function has at least one point $T \in [t, s]$, where $g(T) = 0$, $g'(T) \geq 0$ (as example we take $T = \min\{\tau \geq t \,|\, g(\tau) = 0\}$). It comes back to $F(x(T)) = 0$ and $(\nabla F(x(T)), X(x(T))) \geq 0$, contradiction. □

2.4 Dissipated quantities

Let us consider again an autonomous differential system (dynamical system) $\dot{x}(t) = X(x(t))$, where X is a C^1 vector field on $\mathbb{R}^n$. A C^1 function $F : \mathbb{R}^n \to \mathbb{R}$ is called *dissipated quantity (function)* with respect to the dynamical system if for each trajectory $x(t)$ the composed function $F(x(t))$ is decreasing, i.e., $F(x(\tau)) \leq F(x(t)), \forall \tau \geq t$ or $\frac{d}{dt}F(x(t)) = (\nabla F(x(t)), X(x(t))) \leq 0$. In other words, F is a dissipated quantity if and only if the angle between the vector fields ∇F and X is obtuse.

The classical examples are: (1) the total energy of a dynamic system with damping, (2) the total fluid in an open flowing system.

If $F(x)$ is a dissipated function, then: (1) the constant sublevel sets $N_c : F(x) \leq c$ are invariant; (2) it exists $\lim_{t\to\infty} F(x(t)) = F^*$, where $F^* \in R \cup \{-\infty\}$; (3) if the trajectory $x(t)$ is bounded and $\frac{d}{dt}F(x(t))$ is a continuous function, then the trajectory $x(t)$ converges to the zero-dissipation set

$$x(t) \to \mathcal{D}_0 = \{x \,|\, (\nabla F(x), X(x)) = 0\},$$

i.e., $\lim_{t\to\infty} d(x(t), \mathcal{D}_0) \to 0$.

Consider an autonomous linear system with constant coefficients $\dot{x}(t) = Ax(t)$. Then: (1) a linear function $F(x) = a^T x$ is a dissipated function if and only if $a^T Ax(t) \leq 0$, $\forall x(t)$, i.e., $A^T a = 0$, i.e., the function F is a first integral; (2) a quadratic form $F(x) = x^T Px$, with P a symmetric matrix, satisfies

$$(\nabla F(x), X(x)) = x^T (A^T P + PA)x$$

and hence the quadratic form F is: (i) a first integral (conserved quantity) if and only if $A^T P + PA = 0$; (ii) a dissipated quantity if and only if $A^T P + PA \leq 0$ (negative semidefinite matrix).

2.5 The example of linear mechanical system

A linear mechanical system with damping is described by a second order ODE

$$M\ddot{q}(t) + D\dot{q}(t) + Kq(t) = 0,$$

where $q(t) \in \mathbb{R}^n$ is the displacement or configuration, $M = M^T > 0$ is the mass or inertia matrix (positive definite), $K = K^T > 0$ is the stiffness matrix (positive definite) and $D = D^T \geq 0$ is the damping or loss matrix (positive semidefinite).

If we introduce the notation $x = {}^T(x^1, x^2) = {}^T(q, \dot{q})$, then we can write (matrix form)

$$\dot{x} = \begin{bmatrix} \dot{q} \\ \ddot{q} \end{bmatrix} = \begin{bmatrix} 0 & I \\ -M^{-1}K & -M^{-1}D \end{bmatrix} x.$$

In this context, the total energy of the system (kinetic energy + potential energy) is

$$E = \frac{1}{2}\dot{q}^T M\dot{q} + \frac{1}{2}q^T Kq = \frac{1}{2}x^T \begin{bmatrix} K & 0 \\ 0 & M \end{bmatrix} x.$$

This total energy dissipates because

$$\begin{aligned} \frac{d}{dt}E &= x^T \begin{bmatrix} K & 0 \\ 0 & M \end{bmatrix} \begin{bmatrix} 0 & I \\ -M^{-1}K & -M^{-1}D \end{bmatrix} x \\ &= x^T \begin{bmatrix} 0 & K \\ -K & -D \end{bmatrix} x = -\dot{q}^T D\dot{q} \leq 0. \end{aligned}$$

Physically: $\frac{d}{dt}$ (total energy) = - (dissipated energy).

2.6 Discrete time dynamical systems

A dynamical system of the form $x(t+1) = X(x(t))$, $t \in \mathbb{N}$, is called *discrete time dynamical system* or *recursive system.* The set $S \subset \mathbb{R}^n$ is called *invariant* (with respect to the dynamical system) if for each trajectory $x(t)$ we have

$$x(t) \in S \Rightarrow x(\tau) \in S, \ \forall \tau \geq t,$$

that is, the trajectories can enter the set S but cannot leave S. In other words, $x \in S \Rightarrow X(x) \in T_xS \simeq S$.

A concrete example: the nonnegative orthant $\mathbb{R}^n_+$ is invariant for the discrete dynamical system $x(t+1) = Ax(t),\ A = (a_{ij})$, if and only if $a_{ij} \geq 0,\ i, j = 1, ..., n$.

A function $F : \mathbb{R}^n \to \mathbb{R}$ is conserved in relation to the dynamical system if $F(x(t))$ is a constant for each trajectory $x(t)$, i.e, $F(x(t+1)) = F(x(t))$, for each trajectory $x(t)$. In other words, $F(X(x)) = F(x), \forall x$. A function $F : \mathbb{R}^n \to \mathbb{R}$ is a dissipated quantity if $F(x(t))$ is decreasing for each trajectory $x(t)$, i.e., $F(X(x)) \leq F(x), \forall x$. Sometimes, these notions are formulated with difference $\Delta F(x) = F(X(x)) - F(x)$: (1) the function F is conserved if and only if $\Delta F(x) = 0, \forall x$; (2) the function F is dissipated if and only if $\Delta F(x) \leq 0, \forall x$.

We consider again the discrete dynamical system $x(t+1) = X(x(t))$ and the quadratic form $F(x) = x^T Px$. Since $\Delta F(x) = x^T(A^T PA - P)x$, the quadratic form is conserved if and only if $A^T PA - P = 0$ or dissipated if and only if $A^T PA - P \leq 0$ (negative semidefinite matrix).

2.7 Poincaré section and map

A C^∞ vector field X defines the dynamical system $\dot{x}(t) = X(x(t))$. We denote the associated flow by $T^t(x) = \Phi(t, x)$.

To visualize the flow in the phase space of dimension at least two, H. Poincaré adopted a special point of view. Now this point of view refers to: (i) the *Poincaré section*, a hypersurface in phase space, which is transversal to almost all orbits, (ii) the *Poincaré map*, a function from the Poincaré section to itself (it joins two consecutive intersection points).

Obviously, only the intersection points that come from the same part of the Poincaré section are counted. A Poincaré map transforms a continuous dynamic system into a discrete one. If the Poincaré section is conveniently chosen, then no information is lost from the qualitative behavior of the dynamics. Poincaré maps are reversible because we can find u_n from u_{n+1} following the backword orbits.

Let $F : \mathbb{R}^n \to \mathbb{R}$ be a function of class C^k whose gradient ∇F does not vanish on the subset $\Sigma : F(x) = 0,\ x \in \mathbb{R}^n$. Then, the subset Σ is a *hypersurface*. The hypersurface Σ is called *transversal* to the vector field X, of class C^∞, if $D_X F = (\nabla F, X) \neq 0$.

Theorem 2.7.1. *We fix the initial conditions $x \in \mathbb{R}^n$ and $t_0 \in I_x$.*

Suppose that Σ is a transversal hypersurface to X which goes through the point $T^{t_0}(x)$. Then there exist a neighbourhood U of the point x and a function $\tau(x) \in C^k(U)$ which satisfy the conditions $\tau(x) = t_0$ and $T^{\tau(y)}(y) \in \Sigma$, $\forall y \in U$.

Proof. Let us use the hypersurface $F(T^t(y)) = 0$. It contains at least the point $(t_0, x) \in I \times U$. We compute

$$\frac{\partial F}{\partial t}(T^t(y)) = \frac{\partial F}{\partial x}(T^t(y))X(T^t(y)) \neq 0, \, \forall (t, y) \in I \times U.$$

According to the implicit function theorem, there exists a function $\tau(y) \in C^k(U)$ such that for any $y \in U$ one has $F(T^{\tau(y)}(y)) = 0$, i.e., $T^{\tau(y)}(y) \in \Sigma$, $\forall y \in U$. □

If the current $T^t(x)$ is periodic and if $t_0 = t(x)$ is the period, then the function $P_\Sigma(y) = T^{\tau(y)}(y)$ is called the *Poincaré map*. This transformation applies the hypersurface Σ into itself and each fixed point of the vector field corresponds to a periodic orbit.

2.7.1 Partial derivatives of the Poincaré map

Let us compute the first partial derivatives and the second order partial derivatives of the Poincaré map using the flow and an idea in [6]. The partial function $t \to \Phi(t, x)$ is just an alternative expression for the solution of the Cauchy problem $\frac{\partial}{\partial t}\Phi(t, x) = X(\Phi(t, x))$, $\Phi(0, x) = x$. It follows $\Phi(t, x) = x + \int_0^t X(\Phi(s, x))ds$. Taking the partial derivative with respect to x, we obtain

$$\frac{\partial \Phi}{\partial x}(t, x) = I + \int_0^t \langle \frac{\partial X}{\partial x}(\Phi(s, x)), \frac{\partial \Phi}{\partial x}(s, x)\rangle ds.$$

Denoting $z(t) = \frac{\partial \Phi}{\partial x}(t, x)$, we find $z(0) = \frac{\partial \Phi}{\partial x}(0, x) = I$. Differentiating z with respect to t, we get the Cauchy problem $\dot{z}(t) = \langle \frac{\partial X}{\partial x}(\Phi(t, x)), z(t)\rangle$, $z(0) = I$. Suppose the matrices $A(t) = (\frac{\partial X}{\partial x}(\Phi(t, x)))$ and $\int_0^t A(s)ds$ commute. Then, it follows the solution

$$z(t) = \exp \int_0^t \frac{\partial X}{\partial x}(\Phi(s, x))ds$$

and consequently $\frac{\partial \Phi}{\partial x}(1, x) = \exp \int_0^1 \frac{\partial X}{\partial x}(\Phi(s, x))ds$. The Poincaré map is $P(x) = \Phi(1, x)$ and we have determined the first order partial derivatives $\frac{\partial P}{\partial x} = \exp \int_0^1 \frac{\partial X}{\partial x}(\Phi(s, x))ds$. It follows that $P(x)$ is a partially increasing function. Differentiating once more, we find the second order partial derivatives

$$P_{xx} = P_x \left(\int_0^1 X_{xx}(\Phi(s,x)) \cdot \exp\left(\int_0^s X_x(\Phi(u,x))du \right) ds \right).$$

If $X_{x^i x^j}$ is a positive definite matrix, then $P(x)$ is convex.

The partial derivatives P_x and P_{xx} allow the qualitative study of the Poincaré map $P(x)$ and not its effective determination, since the integrand contains the unknown flow $\Phi(t,x)$.

2.8 Phytoplankton dynamical system

A central problem in biological dynamical systems is to determine the boundaries of evolution. Although this is a general problem, we prefer to give solutions for growth of the phytoplankton [12].

In the paper of Bernard - Gouze [2] is analyzed a model of *phytoplankton growth* based on the *phytoplankton growth vector field* $X = (X^1, X^2, X^3)$ of components

$$X^1 = 1 - x^1 - \frac{1}{4}\,x^1 x^2,\ X^2 = 2x^2 x^3 - x^2,\ X^3 = \frac{1}{4}\,x^1 - 2{x^3}^2,$$

and the dynamical system $\dot{x}^1(t) = X^1(x(t))$, $\dot{x}^2(t) = X^2(x(t))$, $\dot{x}^3(t) = X^3(x(t))$, $x = (x^1, x^2, x^3)$, where x^1 means the substrate, x^2 is the phytoplankton biomass and x^3 is the intracellular nutrient per biomass, with the physical domain $\mathbb{R}^3_+ : x = (x^1,\, x^2,\, x^3) \geq 0$. The previous dynamical system is non-cooperative and has the equilibrium point $\left(1, 0, \frac{1}{2\sqrt{2}}\right)$. We introduce the maximal field line $x = x(t, x_0), t \in I$, which satisfies the initial condition $x(t_0, x_0) = x_0$.

In order to find bounds for substrate, biomass, and intracellular nutrient per biomass, we use the techniques of optimization developed in our papers [6] - [10].

2.8.1 Bounds for phytoplankton substrate

We use the following problem: find $\max f(x^1, x^2, x^3) = x^1$ **with the restriction** $x = x(t, x_0)$.

We set the critical point condition $\langle \nabla f, X \rangle = 0$. In this case $\nabla f = (1, 0, 0)$. It follows the relation $1 - x^1 - \frac{1}{4}\,x^1 x^2 = 0$. The convenient solution (critical point) $\bar{x} = (\bar{x}^1, \bar{x}^2, \bar{x}^3)$ must be

$$\bar{x} = \left(\bar{x}^1 \in [0,1),\ \bar{x}^2 = \frac{4(1-\bar{x}^1)}{\bar{x}^1},\ \bar{x}^3 \geq 0\right).$$

The sufficient condition $Hess\ f(X,X)(\bar{x}) + \langle \nabla f, D_X X\rangle(\bar{x}) < 0$ reduces to $\langle \nabla f, D_X X\rangle(\bar{x}) < 0$ or

$$\left(1+\frac{\bar{x}^2}{4}\right)\left(1-\bar{x}^1-\frac{1}{4}\bar{x}^1\bar{x}^2\right)+\frac{\bar{x}^1}{4}(2\bar{x}^2\bar{x}^3-\bar{x}^2) > 0.$$

Since at the critical point we have $\bar{x}^2 = \frac{4(1-\bar{x}^1)}{\bar{x}^1}$, the condition goes to $(1-\bar{x}^1)(2\bar{x}^3-1) > 0$, and the convenient condition is $\bar{x}^1 < 1,\ 2\bar{x}^3 > 1$.

Theorem 2.8.1. *Suppose that on an evolution line (field line) it exists a point $\bar{x}$ at which we have*

$$\bar{x}^1 < 1,\ 2\bar{x}^3 > 1,\ \bar{x}^2 = \frac{4(1-\bar{x}^1)}{\bar{x}^1}.$$

Then, the phytoplankton substrate has an upper bound at this point.

2.8.2 Bounds for phytoplankton biomass

Let us use the problem: find $\max g(x^1,x^2,x^3) = x^2$ **subject to** $x = x(t, x_0)$.

The critical point condition is $\langle \nabla g, X\rangle = 0$. Since $\nabla g = (0,1,0)$, it follows the relation $2x^2x^3 - x^2 = 0$. The convenient solution (critical point) $\bar{x} = (\bar{x}^1, \bar{x}^2, \bar{x}^3)$ must have $\bar{x}^3 = \frac{1}{2}$. The sufficient condition

$$Hess\ g(X,X)(\bar{x}) + \langle \nabla g, D_X X\rangle(\bar{x}) < 0$$

reduces to $\langle \nabla g, D_X X\rangle(\bar{x}) < 0$ or $\bar{x}^2(2\bar{x}^3-1)^2 + 2\bar{x}^2(\frac{1}{4}\bar{x}^1 - 2\bar{x}^{3^2}) < 0$. Since at the critical point we have $\bar{x}^3 = \frac{1}{2}$, the sufficient condition leads to $\bar{x}^2(\bar{x}^1-2) < 0$, i.e., $\bar{x}^2 > 0,\ \bar{x}^1 - 2 < 0$.

Theorem 2.8.2. *Suppose that on an evolution line (field line) there exists a point $\bar{x}$ at which we have*

$$\bar{x}^1 < 2,\ \bar{x}^2 > 0,\ \bar{x}^3 = \frac{1}{2}.$$

Then, the $x^2(t)$ component of the corresponding field line has an upper bound at this point.

A direct alternative is to build the composite function $g(x(t, x_0))$. The condition

$$\frac{d}{dt}g(x(t,x_0)) = \frac{\partial g}{\partial x^i}(x(t,x_0))\frac{dx^i}{dt}(t) = 0$$

reduces to $\frac{dx^2}{dt}(t) = 0$, i.e., $2x^2x^3 - x^2 = 0$. The convenient solution is $x^3 = \frac{1}{2}$. The condition

$$\frac{d^2}{dt^2}g(x(t,x_0)) = \frac{\partial^2 g}{\partial x^i \partial x^j}(x(t,x_0))\frac{dx^i}{dt}(t)\frac{dx^j}{dt}(t) + \frac{\partial g}{\partial x^i}(x(t,x_0))\frac{d^2x^i}{dt^2}(t) < 0,$$

becomes $2x^2(t)\frac{dx^3}{dt}(t) < 0$, i.e., $x^2 > 0$, $\frac{1}{4}\,x^1 - 2{x^3}^2 < 0$. Replacing $x^3 = \frac{1}{2}$, we find $x^2 > 0$, $x^1 - 2 < 0$. The same result is obtained by using the previous method.

2.8.3 Bounds for intracellular nutrient per biomass

Now the helping problem is: compute max $h(x^1, x^2, x^3) = x^3$ **with the restriction** $x = x(t, x_0)$.

The critical point condition is $\langle \nabla h, X\rangle = 0$. Since $\nabla h = (0, 0, 1)$, it follows the relation $\frac{1}{4}\,x^1 - 2{x^3}^2 = 0$. The critical point $\bar{x} = (\bar{x}^1, \bar{x}^2, \bar{x}^3)$ must be

$$\bar{x} = \left(\bar{x}^1 \geq 0,\ \bar{x}^2 \geq 0,\ \bar{x}^3 = \frac{\sqrt{\bar{x}^1}}{2\sqrt{2}}\right).$$

The sufficient condition $Hess\ h(X, X)(\bar{x}) + \langle \nabla h, D_X X\rangle(\bar{x}) < 0$ reduces to $\langle \nabla h, D_X X\rangle(\bar{x}) < 0$ or

$$\frac{1}{4}\left(1 - \bar{x}^1 - \frac{1}{4}\,\bar{x}^1\bar{x}^2\right) - 4\bar{x}^3\left(\frac{1}{4}\bar{x}^1 - 2{\bar{x}^3}^2\right) < 0.$$

Since at the critical point we have $\bar{x}^3 = \frac{\sqrt{\bar{x}^1}}{2\sqrt{2}}$, the condition goes to $1 - \bar{x}^1 - \frac{1}{4}\,\bar{x}^1\bar{x}^2 < 0$, i.e.,

$$1 < \bar{x}^1\left(1 + \frac{1}{4}\bar{x}^2\right).$$

Theorem 2.8.3. *Suppose that on an evolution line (field line) there exists a point* $\bar{x}$ *at which we have*

$$1 < \bar{x}^1\left(1 + \frac{1}{4}\bar{x}^2\right),\ \bar{x}^3 = \frac{\sqrt{\bar{x}^1}}{2\sqrt{2}}.$$

Then, the intracellular nutrient per biomass has an upper bound at this point.

2.9 Research problems

2.9.1 Invariant sets, conservation laws, dissipation

Problem 8. *For $x = (x^1, ..., x^n) \in \mathbb{R}^n$, we write $x > 0$ if $x^i > 0$ and $x \geq 0$ if $x^i \geq 0$. The set $\mathbb{R}^n_+ = \{x \in \mathbb{R}^n : x \geq 0\}$ is called positive orthant. Most of the time we use the open positive orthant $\mathbb{R}^n_{++} = \{x \in \mathbb{R}^n : x > 0\}$. On the set $\mathbb{R}^n_{++}$, let us take the distinct monomials of the form $v^k = (x^1)^{\alpha_{1k}} \cdots (x^n)^{\alpha_{nk}}$, $k = 1, ..., m$, where α_{ik} are real numbers. If a^i_k are real numbers, then the functions $a^i_k v^k$, summed by the index k, are called sygnomials. The sygnomial dynamical systems are defined by*

$$\dot{x}^i(t) = a^i_k v^k(x(t)).$$

Using the change of variables $y^i = \ln x^i$, show the sygnomial ODE system rewrites $\dot{y} = \left(Ae^{By}\right) e^{-y}$ or $\dot{x}(t) = x(t) \otimes \left(Ae^{B \ln x(t)}\right) e^{-\ln x(t)}$, where $A = (a^i_k)$, $B = (\alpha_{ki})$. Discuss the existence of equilibrium points. Show that if $r = rank(A)$, then the ODE system in y has $n - r$ first integrals of type $a^T y = const.$

Problem 9. *We consider the model of phytoplankton growth*

$$\dot{x}^1 = 1 - x^1 - \frac{1}{4}x^1x^2, \; \dot{x}^2 = 2x^2x^3 - x^2, \; \dot{x}^3 = \frac{1}{4}x^1 - 2{x^3}^2,$$

were x^1 means the substrate, x^2 is the biomass of phytoplankton and x^3 is the intercellular nutrient per biomass. Find: equilibrium points, an invariant set that does not contain equilibrium points, a conservation law and a dissipation law.

Solution. The invariance of a parallelepiped from the first orthant is related to the existence of opposite diagonal points $l = (l^1, l^2, l^3)$ and $u = (u^1, u^2, u^3)$ which satisfy

$$1 - u^1 - \frac{1}{4}u^1 l^2 \leq 0, \; u^2(2u^3 - 1) \leq 0, \; \frac{1}{4}u^1 - 2{u^3}^2 \leq 0$$

$$1 - l^1 - \frac{1}{4}l^1 u^2 \leq 0, \; l^2(2l^3 - 1) \leq 0, \; \frac{1}{4}l^1 - 2{l^3}^2 \leq 0.$$

A possible solution is $l = (0, 0, 0)$ and $u = \left(2, 1, \frac{1}{2}\right)$. Consequently, the parallelepiped with the diagonally opposite ends l and u is an invariant set in the first orthant.

Problem 10. *Let x^1, x^2, x^3, x^4, respectively, cytoplasmic concentrations of protein $TetR$, complex $TetR - Tc$, tetracycline and protein θ. Suppose that u_0 is the extracellular concentration of tetracycline. Let f be the rate of transcription of genes that are inhibited by $TetR$. Dynamics of tetracycline antibiotic in a bacterium that develops drug resistance (switching between the $TetA$ and $TetR$ genes) can be described by the hybrid system*

$$\dot{x}^1 = f - \frac{1}{3}\,x^1x^3 + \frac{1}{8000}\,x^2,\ \dot{x}^2 = \frac{3}{200}\,u_0 - \frac{7}{2}\,x^3x^4$$

$$\dot{x}^3 = \frac{1}{3}\,x^1x^3 - \frac{1}{2500}\,x^2,\ \dot{x}^4 = f - \frac{11}{40000}\,x^4,$$

where $f = \frac{1}{2000}$ if $TetR > \frac{1}{50000}$ and $f = \frac{1}{40}$ otherwise. Find: equilibrium points, another invariant set, a law of conservation and a law of dissipation.

Problem 11. *Let $x^i, i = 1,...,n$ be the state variables (metabolic concentrations, protein concentrations, gene expression levels etc). Let b^i, a^i_α, $\alpha = 1,...,m$, stoichiometric coefficients, and $\sigma_{\alpha j}$, $j = 1,...,n$, be kinetic orders. Some of the dynamic biological systems are defined by sygnomial vector fields. They are shaped*

$$\dot{x}^i(t) = b^i x^i(t) + x^i(t)\sum_{\alpha=1}^{m} a^i_\alpha \prod_{j=1}^{n}(x^j(t))^{\sigma_{\alpha j}};\ i,j = 1,...,n; \alpha = 1,...,m.$$

One example is Lotka-Volterra dynamical system in several dimensions

$$\dot{z}^i(t) = \lambda^i z^i(t)\left(1 - a^i_j z^j(t)\right).$$

Problem 12. *(i) Find a first integral (conservation law) for the dynamical system*

$$\dot{x}_i = 2\left(c_i - \sum_{j=1}^{n} c_j x_j\right) x_i,\ x_i \geq 0,\ \sum_{i=1}^{n} x_i = 1,\ i,j = 1,...,n.$$

(ii) Find a conservation law for the dynamical system

$$\dot{x}_i = 2\left(c_i x_i - \sum_{j=1}^{n} c_j x_j^2\right) x_i,\ x_i \geq 0,\ \sum_{i=1}^{n} x_i = 1;\ i,j = 1,...,n.$$

(iii) Deduce a conservation law for the dynamical system

$$\dot{x}_i = 2\left(c_i f_i(x_i) - \sum_{j=1}^{n} c_j f_j(x_j) x_j\right) x_i;\ i,j = 1, ..., n,$$

where $f_j : [0,1] \to R$ are C^1 monotone increasing functions.

Hint $\sum_{i=1}^{n} \dot{x}_i = 0$. (i) The solution is

$$x(t) = \frac{e^{2tN} x(0)}{\langle e^{2tN} x(0)\rangle},$$

where $N = \text{diag}(c_1, ..., c_n)$ and $\langle e^{2tN} x(0)\rangle = \sum_{j=1}^{n} e^{2tc_j} x_j(0)$.

Problem 13. *On the Riemannian manifold $(\mathbb{R}, g(x))$, analyse the "minus gradient" flow*

$$\dot{x}(t) + \frac{1}{g(x(t))} f'(x(t)) = 0.$$

For $g(x) = 1 + x^2$ and $f(x) = x^3$, detailed discussion.

2.10 Self-evaluation problems

2.10.1 Energy conservation and dissipation

Problem 14 (Harmonic pendulum with friction)**.** *The motion of harmonic pendulum with friction is modelled by the dynamical system*

$$\dot{x} = v,\ \dot{v} = -\frac{g}{L} x - \gamma v,\ \gamma > 0.$$

Show that the associated flow is contractive.

Hint The vector field $X = (v, -\frac{g}{L} x - \gamma v)$ has the divergence $\nabla X = -\gamma < 0$. Consequently, the flow is contractive.

Problem 15. *The inertial particle motion is described by the dynamical system*

$$\frac{dx}{dt} = v,\ \frac{dv}{dt} = \beta \frac{\partial}{\partial t} u(x,t) + \frac{u(x,t) - v}{\tau_p}.$$

Show that the dynamics is dissipative.

Hint Let $St = \tau_p$ be the Stokes number. The vector field $X = (v, \beta\, \frac{\partial}{\partial t} u(x,t) + \frac{u(x,t)-v}{St})$ has a Jacobian matrix of elements $\frac{\partial X^i}{\partial \xi^j} = A^i_j$. Denote $\sigma_{ij} = \partial_j u_i$. The linearization of the dynamical system has the matrix

$$A = \begin{pmatrix} 0 & I \\ \beta\partial_t \sigma + \frac{\sigma}{\tau_p} & -\frac{I}{\tau_p} \end{pmatrix}.$$

It follows the divergence $\nabla X = trA = -\dfrac{3}{\tau_p} < 0$ (uniform contraction).

Problem 16. *Let $m\ddot{q}(t) = -F(q(t))$ be a dynamical system, where $m > 0$ is the mass, $q(t)$ is the displacement and $F(0) = 0$. Show that the total energy $E = \dfrac{m}{2}\,\dot{q}^2 + \displaystyle\int_0^q F(u)du$ is conserved.*

Solution. We compute the total derivative of energy (with respect to the time variable), taking into account the equation that describes the dynamical system. It follows

$$\dot{E}(q(t), \dot{q}(t)) = m\dot{q}(t)\ddot{q}(t) + \frac{d}{dt}\left(\int_0^{q(t)} F(u)dt\right)$$

$$= m\dot{q}(t)\ddot{q}(t) + \dot{q}(t)F(q(t)) = \dot{q}(t)(m\ddot{q}(t) + F(q(t)) = 0.$$

Thus the energy $E(q, \dot{q})$ is constant as a function of time, so in this case the energy is conserved.

Problem 17. *Let $M\ddot{q} + D\dot{q} + Kq = 0$ be a linear mechanical system with damping, where $M = M^T$ is the mass matrix, positive definite, $K = K^T$ is the matrix of rigidity, positive definite, and $D = D^T$ is the damping matrix, positive semidefinite. Show that the total energy $E = \dfrac{1}{2}\dot{q}^T M\dot{q} + \dfrac{1}{2}q^T Kq$ is dissipative.*

Solution. Compute the total energy derivative (with respect to the time variable) using the equation that describes the mechanical system, the symmetry of the matrices M, K, D, as well as the properties of the matrix transposition operation. Deduce

$$\dot{E}(q(t), \dot{q}(t)) = \frac{1}{2}\ddot{q}(t)^T M\dot{q}(t) + \frac{1}{2}\dot{q}(t)^T M\ddot{q}(t) + \frac{1}{2}\dot{q}(t)^T Kq(t) + \frac{1}{2}q(t)^T K\dot{q}(t)$$

$$= \frac{1}{2}\left(\ddot{q}(t)^T M + q(t)^T K\right)\dot{q}(t) + \frac{1}{2}\dot{q}(t)^T\left(M\ddot{q}(t) + Kq(t)\right)$$

$$= \frac{1}{2}\left(M^T\ddot{q}(t) + K^T q(t)\right)^T \dot{q}(t) - \frac{1}{2}\dot{q}(t)^T D\dot{q}(t)$$

$$= \frac{1}{2}(-D\dot{q}(t))^T\dot{q}(t) - \frac{1}{2}\dot{q}(t)^T D\dot{q}(t) = -\dot{q}(t)^T D\dot{q}(t).$$

But D is a positive semidefinite matrix, i.e., $X^T DX \geq 0, \ \forall X$. Then $\dot{E}(q(t), \dot{q}(t)) \leq 0$, which means that the total energy E is dissipative.

Problem 18. *Show that (up to an overall factor) the characteristic polynomial for the linear mechanical system with damping is*

$$P(\lambda) = \det[\lambda^2 M + \lambda D + K]$$

and that this actually is a polynomial of degree n in λ^2.

Problem 19. *Let $\dot{x}(t) = Ax(t)$ be a linear constant coefficients dynamical system. Find the conditions so that the quadratic form $F(x) = x^T Px$ be conservative or dissipative.*

Solution. The derivative of the composed function $F(x(t))$ with respect to time variable is

$$\dot{F}(x(t)) = \dot{x}(t)^T Px(t) + x(t)^T P\dot{x}(t) = (Ax(t))^T Px(t) + x(t)^T P(Ax(t))$$

$$= x(t)^T A^T Px(t) + x(t)^T PAx(t) = x(t)^T (A^T P + PA)x(t).$$

The quadratic form F is conservative if and only if $\dot{F}(x(t)) = 0, \ \forall x(t)$, i.e., $A^T P + PA = 0$.

The quadratic form F is dissipative if and only if $\dot{F}(x(t)) \leq 0, \ \forall x(t)$, i.e, $x(t)^T (A^T P + PA)x(t) \leq 0, \ \forall x(t)$, which implies that the matrix $A^T P + PA$ is negative semidefinite.

Bibliography

[1] V. I. Arnold, *Geometrical Methods in the Theory of Ordinary Differential Equations*, Springer-Verlag, New York, 1983.

[2] O. Bernard, J.-L. Gouze, *Global qualitative description of a class of nonlinear dynamical systems*, Artificial Intelligence, 136, (2002), pp. 29-59.

[3] O. Calin, C. Udriste, *Geometric Modeling in Probability and Statistics*, Springer, 2014.

[4] Cours EE363, Lecture 9, *Invariant Sets, Conservation, and Dissipation*, INTERNET, 2011.

[5] E. Hairer, C. Lubich, G. Wanner, *Geometric Numerical Integration*, Springer, 2002.

[6] M. W. Hirsch, S. Smale, R. L . Devaney, *Differential Equations, Dynamical Systems and Introduction to Chaos*, Academic Press, An imprint of Elsevier, 2004.

[7] C.-G. Lefter, *Differential Equation and Dynamical Systems* (in Romanian), Alexandru Myller Editorial House, Iaşi, 2006.

[8] C. Udriste, *Convex Functions and Optimization Methods on Riemannian Manifolds,* Mathematics and Its Aplications 297, Kluwer Academic Publishers, 1994.

[9] C. Udriste, *Geometric Dynamics*, Mathematics and Its Applications, 513, Kluwer Academic Publishers, Dordrecht, Boston, London, 2000.

[10] C. Udriste, V. Damian, *Simplified single-time stochastic maximum principle*, Balk. J. Geom. Appl., 16, 2 (2011), pp. 155-173.

[11] C. Udriste, O. Dogaru, *Extrema conditioned by orbits* (in Romanian), Sci. Bull. IPB, 51 (1991), 3-9.

[12] C. Udriste, O. Dogaru, I. Tevy, *Extrema with Nonholonomic Constraints*, Geometry Balkan Press, 2002.

[13] C. Udriste, L. Matei, *Lagrange-Hamilton Theories* (in Romanian), Monographs and Textbooks 8, Geometry Balkan Press, Bucharest, 2008.

[14] C. Udriste, I. Tevy, *Growth of Phytoplankton*, Biomed J Sci & Tech Res (BJSTR), 13, 2 (2019), 4793-4794, DOI: 10.26717/BJSTR.2019.13.002360.

[15] I. I. Vrabie, *Differential Equation* (in Romanian), MATRIXROM Editorial House, Bucharest, 2000.

[16] E. T. Whittaker, *A Treatise on The Analytical Dynamics of Particles & Rigid Bodies*, Cambridge University Press, 1989.

Chapter 3

EQUILIBRIUM POINTS STABILITY AND HOPF BIFURCATION

Motto:
"Those who want to fight live;
And the scared ones complain and die.
If you see them die, let them die,
Because death is their destiny."

G. Coşbuc - *Life's struggle*

In this Chapter, we analyze two problems: the stability of the equilibrium positions and the Hopf bifurcation of a flow. Local stability refers to the stability of the equilibrium points of the linearized system. Global stability is obtained from the use of Lyapunov functions. Hopf bifurcation refers to the possibility of switching from closed trajectories to open trajectories. Such problems have direct applications in science and engineering, being related to phenomena of evolution over time. Suggestions for further reading: [1]-[8].

3.1 Local stability of equilibrium points

Let X be a C^∞ vector field on $\mathbb{R}^n$ which determines the dynamical system $\dot{x}(t) = X(x(t))$, $t \in I \subset \mathbb{R}$. The solutions $x = x(t)$ of this differential system are called *field lines* of X, and their images in $\mathbb{R}^n$ are called *orbits*. Regarding the solutions of this system of differential equations, we recall the theorem of existence and uniqueness: if X is local Lipschitz, then the solution of the Cauchy problem

$$\dot{x}(t) = X(x(t)),\, x(0) = x_0$$

there exists and is unique. A solution $x = a$ of the algebraic system $X(x) = 0$ generates a constant solution $x(t) = a$ of the dynamical system, called *equilibrium point.*

Suppose that $x(t, \epsilon)$, $\epsilon \in (-\delta, \delta)$ is a smooth variation of a solution $x(t)$, i.e.,

$$\dot{x}(t,\epsilon) = X(x(t,\epsilon)),\, x(t,0) = x(t).$$

Taking the derivative with respect to ϵ, setting $\epsilon = 0$ and denoting $\xi = \frac{\partial x}{\partial \epsilon}|_{\epsilon=0}$, we find the *differential system in variations*

$$\dot{\xi}(t) = \frac{\partial X}{\partial x}(x(t))\, \xi(t),$$

or the *linearized differential system* along a trajectory of the initial system, where $A = \dfrac{\partial X}{\partial x}$ is the Jacobian matrix. If the trajectory is reduced to an equilibrium point (constant trajectory), for example $x(t) = 0$, then the linear approximation around this point is usually written in matriceal notation

$$\dot{x}(t) = A\, x(t),\, A = \frac{\partial X}{\partial x}(0).$$

Theorem 3.1.1. *Suppose that the equilibrium point $x(t) = 0$ of the dynamical system $\dot{x}(t) = A\, x(t)$ is isolated.*

(1) If all the eigenvalues of the matrix A have negative real part, then the equilibrium position is asymptotically stable (and therefore stable).

(2) If the eigenvalues of the matrix A are strictly negative (real) or complex with the negative real part, and the purely imaginary eigenvalues exist with the same algebraic and geometric multiplicity, then the equilibrium position is stable but not asymptotically stable.

(3) If an eigenvalue of the matrix A has a positive real part or if there is an imaginary eigenvalue such that the geometric multiplicity of this eigenvalue is less than its algebraic multiplicity, then the equilibrium position is not stable.

The following theorem shows that the asymptotic stability of the linear approximation implies the asymptotic stability for the nonlinear system.

Theorem 3.1.2. *We consider a dynamical system $\dot{x}(t) = X(x(t))$ with isolated equilibrium point $x(t) = 0$ and denote $A = \frac{\partial X}{\partial x}(0)$.*

(1) If all the eigenvalues of the matrix A have strictly negative real part, then the equilibrium point is asymptotically stable (and hence stable).

(2) If the matrix A has one eigenvalue with strictly positive real part, then the equilibrium point is unstable.

3.2 Global stability of equilibrium points

Let X be a vector field of class C^∞ on $\mathbb{R}^n$ which determines the dynamical system $\dot{x}(t) = X(x(t))$. Let $f : \mathbb{R}^n \to \mathbb{R}$ be a scalar field of class C^1. We consider the derivative $D_X f = \langle \nabla f, X \rangle$. Along each solution of the dynamical system we have

$$\frac{d}{dt} f(x(t)) = \langle \nabla f(x(t)), X(x(t)) \rangle = D_X f(x(t)),$$

where the derivative operator D_X is called *derivative along field lines of X.*

The scalar field (function) f is called: (1) *positive definite* if $f(0) = 0$ and $f(x) > 0$, $\forall x \in \mathbb{R}^n \setminus \{0\}$; (2) *positive semidefinite* if $f(0) = 0$ and $f(x) \geq 0$, $\forall x \in \mathbb{R}^n \setminus \{0\}$.

The function (scalar field) f is called *Lyapunov function of stability* if it is positive definite and the function $D_X f$ is negative semidefinite on a neighbourhood V of origin. The function f is called *Lyapunov function of unstability* if it is positive definite and if the function $D_X f$ is positive definite on a neighbourhood V of origin.

Suppose that the dynamical system $\dot{x}(t) = X(x(t))$ admits the isolated equilibrium point $x(t) = 0$. Then, the following three Theorems hold.

Theorem 3.2.1. *If there exists a Lyapunov function f of stability, then the equilibrium point $x = 0$ is stable.*

Theorem 3.2.2. *If there exists a Lyapunov function f of stability and if the set*

$$M_0 = \{x \,|\, x \in V,\ D_X f(x) = 0\}$$

does not contain entirely any non-constant orbit of the dynamical system, then the equilibrium point $x = 0$ is asymptotically stable.

Theorem 3.2.3. *If there exists a Lyapunov function f of unstability, then the equilibrium point $x = 0$ is unstable.*

For a second order differential system, a Lyapunov function is a scalar function defined on the phase space, which can be used to prove the stability of an equilibrium point. The Lyapunov function method is applied to study the stability of various differential equations and systems. This idea can be illustrated with the familiar example of a damped harmonic oscillator. A spring mass system with linear damping is modelled by the second order differential equation

$$m\,\ddot{y}(t) + c\,\dot{y}(t) + k\,y(t) = 0.$$

The total energy of such a system is

$$E(y(t), \dot{y}(t)) = \frac{1}{2}\,k\,y^2(t) + \frac{1}{2}\,m\,\dot{y}^2(t).$$

In the phase plane, the constant level curves $E(y, \dot{y}) = c$ represent concentric ellipses around the equilibrium point $y = 0, \dot{y} = 0$. By computation, we find

$$\frac{d}{dt}E(y(t), \dot{y}(t)) = -c\,\dot{y}^2(t),$$

and hence the trajectories are moving inward across the equal energy contours represented by ellipses. On this basis, the stability can be concluded.

3.3 Hopf bifurcation theory

Let $X(x, c)$ be a vector field of class C^∞ on $\mathbb{R}^n$ which depends on the real parameter $c \in I \subset \mathbb{R}$ and let

$$\dot{x}(t) = X(x(t), c)$$

be the associated dynamical system. Suppose that

$$x = x(c) = (x^1(c), ..., x^n(c))$$

is an isolated equilibrium point, for a fixed value of the parameter c. We introduce the Jacobian matrix $A(c) = \left(\frac{\partial X^i}{\partial x^j}\right)(x(c), c)$ and the linear approximation

$$\frac{dx}{dt}(t) = A(c)x(t)$$

around the equilibrium point $x(c)$.

Theorem 3.3.1 (Hopf Theorem). *We denote by $\lambda_1(c), ..., \lambda_n(c)$ the eigenvalues (real or complex numbers) of the matrix $A(c)$. Suppose*

$$\lambda_1(c) = \alpha(c) + i\beta(c),\ \lambda_2(c) = \alpha(c) - i\beta(c)$$

(complex conjugate numbers). If $n > 2$, we also add the conditions

$$\Re\lambda_3(c) < 0, ..., \Re\lambda_n(c) < 0.$$

We assume that there is an isolated value $c_0 \in I$ such that

$$\alpha(c_0) = 0,\ \beta(c_0) \neq 0,\ \frac{d\alpha}{dc}(c_0) > 0$$

(i.e., when c passes through c_0, the function $c \to \alpha(c)$ goes from negative to positive values; the opposite inequality can be also used).

In the previous hypotheses, the following statement (1) and one of statements (2) and (3) are true:

(1) The equilibrium point is a center, that is, there is an infinite set of concentric closed orbits around the point $x_0 = x(c_0)$. If $c \neq c_0$, neighbour to c_0, then there is no periodic orbit around the point $x(c)$.

(2) There is a number $b > c_0$ so that for each $c \in (c_0, b)$ there is a closed orbit and only one around the point of equilibrium $x(c)$, in a neighbourhood of this point. This family of closed orbits, with one parameter, bifurcates at the equilibrium point $x(c_0)$ in the sense that, if $c \to c_0$, then the diameter of the closed orbits varies with $|c - c_0|^{\frac{1}{2}}$. In this case, for $c \leq c_0$, $c \in I$, there are no closed orbits in a neighbourhood of the point $x(c)$.

(3) There is a number $a < c_0$ so that for every $c \in (a, c_0)$ there is a closed orbit and only one around the point of equilibrium $x(c)$, in a neighbourhood of this point. This family of closed orbits, with

one parameter, bifurcates at the equilibrium point $x(c_0)$ in the sense that, if $c \to c_0$, then the diameter of the closed orbits varies with $|c - c_0|^{\frac{1}{2}}$. In this case, for $c \geq c_0$, $c \in I$, there are no closed orbits in a neighbourhood of the point $x(c)$.

The bifurcation of the periodic current (meaning the passage from closed or periodic to open or non-periodic orbits) occurs only for dynamic systems defined on spaces with dimension $n \geq 2$. Recent proofs of this theorem exploit the fact that under the Theorem, in the space of phases, there is a 2-dimensional submanifold that contains the equilibrium point and which plays the role of the eigen - plane from the linear differential systems associated with the eigenvalues λ_1 and λ_2. This submanifold is called the *center manifold.* So the Hopf bifurcations are at least 2-dimensional phenomena.

In some teaching materials, a variant of Hopf's Theorem is presented.

Theorem 3.3.2. *Suppose that $X(0,c) = 0, \forall c \in I$. Let c_0 be a bifurcation point, i.e., for $c < c_0$, the trivial solution is asymptotically stable, and for $c > c_0$ it is unstable. If for $c = c_0$ there exists the eigenvalues $\lambda_1 = \bar{\lambda}_2$ with $\Re\lambda_1|_{c=c_0} = 0$, $\frac{d}{dc}\Re\lambda_1|_{c=c_0} > 0$, and all the others eigenvalues have negative real parts, then for $c > c_0$ there exists a limit cycle with the period $T = \frac{2\pi}{\Re\lambda_1}$. In addition, if the trivial solution is asymptotically stable for $c = c_0$, then the limit cycle is attractive.*

A Hopf Bifurcation occurs when a periodic solution or limit cycle, surrounding an equilibrium point, arises or goes away as a parameter c varies. When a stable limit cycle surrounds an unstable equilibrium point, the bifurcation is called a *supercritical Hopf bifurcation.* If the limit cycle is unstable and surrounds a stable equilibrium point, then the bifurcation is called a *subcritical Hopf bifurcation.*

Example 3.3.3. [1] Investigate the effect of changing the parameter c on the dynamical system

$$\dot{x}(t) = cx(t) - y(t) + \left(x(t) + \frac{3}{2}\, y(t)\right)(x^2(t) + y^2(t)),$$

$$\dot{y}(t) = x(t) + cy(t) + \left(\frac{3}{2}\, x(t) - y(t)\right)(x^2(t) + y^2(t)).$$

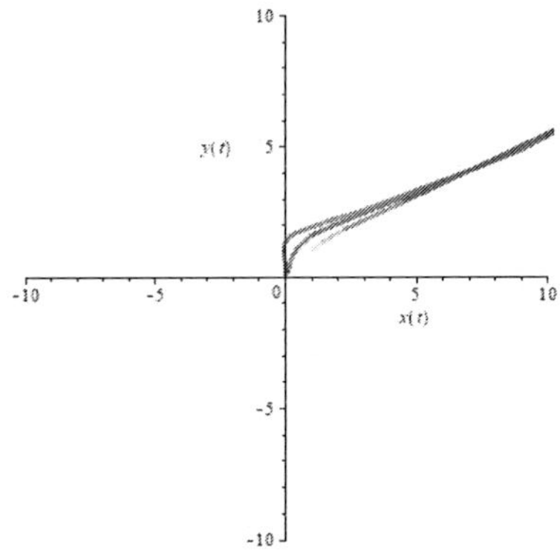

Figure 3.1: Origin is unstable

To simplify, we prefer to pass to the polar form of the system

$$\dot{\rho}(t) = c\rho(t) - \rho^3(t),\ \dot{\theta}(t) = 1 + \frac{3}{2}\,\rho^2(t).$$

from which it is clear that there is a single equilibrium point at the origin. For $c > 0$, the first polar equation can be written as

$$\dot{\rho}(t) = \rho(t)\left(\sqrt{c} + \rho(t)\right)\left(\sqrt{c} - \rho(t)\right).$$

Clearly there is a limit cycle of radius $\sqrt{c}$ in this case. Also the origin is unstable since for small ρ, we have $\dot{\rho} \approx c\rho$. If $c \leq 0$, the $\dot{\rho} < 0$ and the origin is globally asymptotically stable. This behaviour is illustrated in Fig. 1 and Fig. 2, for the cases $c = 4$ and $c = -4$. We see that as the parameter c passes through zero from negative to positive, a stable focus gives way to an unstable focus surrounded by a stable limit cycle whose radius increases with $\sqrt{c}$. This is an example of a Hopf bifurcation for the case $n = 2$.

Maple simulations, Fig. 1 and Fig. 2

> with(DEtools)

> phaseportrait$([(D(x))(t) = 4*x(t) - y(t) + (x(t) + (3/2)*y(t))*(x(t)^2 + y(t)^2), (D(y))(t) = x(t) + 4*y(t) + ((3/2)*x(t) - y(t))*(x(t)^2 + y(t)^2)]$, [x(t), y(t)], t = 0 .. 20, [[x(0) = 0.1e-1, y(0) = 0.1e-1], [x(0) = 0.6e-1, y(0) = 0.6e-1], [x(0) = 1, y(0) = 1]], stepsize = 0.5e-1, scene = [x(t), y(t)], linecolour = sin((1/2)*t*Pi), method = classical[foreuler])

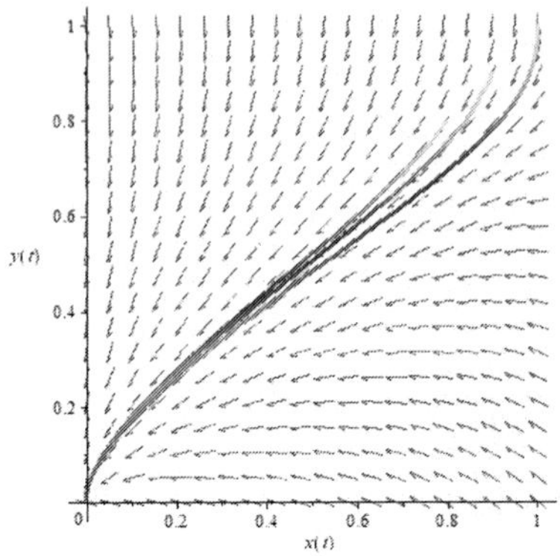

Figure 3.2: Origin is asymptotically stable

> phaseportrait([$(D(x))(t) = -4 * x(t) - y(t) + (x(t) + (3/2) * y(t)) * (x(t)^2 + y(t)^2)$, $(D(y))(t) = x(t) - 4 * y(t) + ((3/2) * x(t) - y(t)) * (x(t)^2 + y(t)^2)$], [x(t), y(t)], t = 0 .. 10, [[x(0) = .8, y(0) = .8], [x(0) = .9, y(0) = .9], [x(0) = 1, y(0) = 1]], stepsize = 0.1e-1, scene = [x(t), y(t)], linecolour = sin((1/2)*t*Pi), method = classical[foreuler])

3.4 Lotka - Voltera dynamical system

The best known model in ecology is the *prey-predator system*

$$\dot{x}(t) = (1 - y(t))x(t), \; \dot{y}(t) = \alpha(x(t) - 1)y(t),$$

introduced by Voltera and Lotka (x is the number of prey - for example, rabbits; y is the number of some predator - for example, foxes). Here α is a positive real parameter describing the interaction of the two species. This system has two equilibrium points: the point $(0, 0)$ which is saddle point, with stable manifold $Oy : x = 0$ and unstable manifold $Ox : y = 0$; the point (1,1) which is not hyperbolic and whose stability cannot be obtained from linearization. The positive quadrant $Q = \{(x, y) \,|\, x > 0, y > 0\}$ is invariant with respect to the Lotka - Voltera flow.

Eliminating the parameter t, by dividing the two equations, we find the equation with separable variables

$$\frac{dy}{dx} = \frac{dy}{dt}\left(\frac{dx}{dt}\right)^{-1} = \alpha\,\frac{(x-1)y}{(1-y)x}.$$

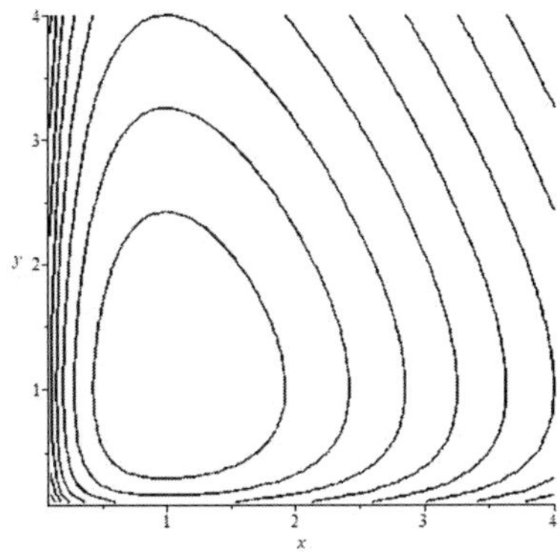

Figure 3.3: Lotka - Volterra orbits

Integrating this ODE, we obtain the orbits

$$f(y) + \alpha f(x) = c, \ f(x) = x - 1 - \ln x, \ x > 0.$$

Obviously, the function f is convex, has the global minimum at the point $x = 1$ and $\lim_{x \to 0} f(x) = \infty$, $\lim_{x \to \infty} f(x) = \infty$. It follows that the constant sublevel sets $f(y) + \alpha f(x) \leq c$ are compact, and the previous orbits are closed (periodic) around the point $(1, 1)$. Thus all the orbits of the Lotka - Volterra system, starting from points in Q, stay in Q (see Fig. 3).

Maple simulation

$>$ with(plots);

$>$ contourplot$(y-1-ln(y)+2*(x-1-ln(x)), x = .1..4, y = .1..4)$;

3.5 Lorenz dynamical system

The autonomous differential system

$$\frac{dx}{dt} = -\sigma x + \sigma y, \ \frac{dy}{dt} = -xz + rx - y, \ \frac{dz}{dt} = xy - bz,$$

where σ, r, b are real parameters, is the mathematical model accepted by Lorenz for the dynamic turbulence of a fluid. More precisely, the fluid evolves between two parallel plates subjected to different temperatures. The evolution is described by the Navier-Stokes equation. Developing in Fourier series according to the spatial coordinates and equating the coefficients with zero, the previous differential system is obtained, where x is proportional to the convective motion, y is proportional to temperature, and z is proportional to the linear temperature distortion of the vertical temperature profile. Usually the following conditions are set: $\sigma > 0, r > 0, b > 0$.

Since $\operatorname{div} X = -(1+\sigma+b) < 0$, the Lorentz flow is contractive.

First, we notice that the Lorenz system is invariant with respect to the symmetry $(x, y, z) \to (-x, -y, z)$. The axis Oz is invariant since $x(t) = 0, y(t) = 0, z(t) = z_0 e^{-bt}, t \in \mathbb{R}$ is a solution of the system.

Let $r < 1$. Then there exists only one equilibrium point $(0, 0, 0)$. The Jacobian matrix, i.e., the matrix of the linear approximation at origin, is

$$\begin{pmatrix} -\sigma & \sigma & 0 \\ r & -1 & 0 \\ 0 & 0 & -b \end{pmatrix}$$

and has three eigenvalues

$$-b, \ -\frac{1}{2}\left(1+\sigma \pm \sqrt{(1+\sigma)^2 + 4(r-1)\sigma}\right).$$

In this way, the origin is an asymptotically stable point. The same thing is confirmed by the Lyapunov function

$$f(x, y, z) = rx^2 + \sigma y^2 + \sigma z^2$$

with the derivative

$$D_X f = -\sigma(r(x+y)^2 + (1-r)y^2 + bz^2),$$

which also applies to the case $r = 1$.

If $r > 1$, then the equilibrium points of the Lorenz system are

$$x = 0, \ y = 0, \ z = 0$$

$$x_0 = \pm\sqrt{b(r-1)}, \ y_0 = \pm\sqrt{b(r-1)}, \ z_0 = r - 1.$$

The origin is no longer a stable point. The linear approximation matrix around the point (x_0, y_0, z_0), is

$$\begin{pmatrix} -\sigma & \sigma & 0 \\ 1 & -1 & -x_0 \\ x_0 & x_0 & -b \end{pmatrix}.$$

For both points, the matrix has the same eigenvalues, but the calculations are laborious. It can be verified that the two equilibrium positions are asymptotically stable for $1 < r < \frac{47}{19} \approx 2.74$.

Let us show that there is an invariant ellipsoid. For this we modify the Lyapunov function above, considering the function

$$F(x, y, z) = rx^2 + \sigma y^2 + \sigma(z - 2r)^2.$$

Using the derivative (quadratic form)

$$D_X F(x, y, z) = -2\sigma \left(r \left(x - \frac{y}{2r} \right)^2 + \frac{3}{r}\, y^2 + b \left(z - \frac{r}{2} \right)^2 \right) + \frac{1}{2}\sigma b r^2,$$

we define the ellipsoid

$$E = \{(x, y, z) \,|\, D_X F(x, y, z) \geq 0\}$$

and the number $M = \max_{(x,y,z)\in E} F(x, y, z)$. We add the ellipsoid

$$\mathcal{E}_\epsilon = \{(x, y, z) \,|\, F(x, y, z) \leq M + \epsilon,\ \epsilon > 0\}.$$

Any point outside the ellipsoid $\mathcal{E}_\epsilon$ is outside of E because

$$D_X F(x, y, z) \leq -\delta < 0$$

for such points. In other words, for $x \in R^3 \setminus E_\epsilon$, the function $F(x(t))$ is strictly decreasing and so the trajectory must enter in the set $\mathcal{E}_\epsilon$ after a finite time. In addition, the ellipsoid $\mathcal{E}_\epsilon$ is a trap region of the Lorenz system and there exists an attractive set $\Lambda = \cap_{n\in\mathbb{N}} \Phi(n, \mathcal{E}_0)$. Particularly, we remark that the field lines exist for $t \geq 0$.

Assuming that σ and b have fixed values, it remains r as a parameter and we investigate the Hopf bifurcation in relation to r. We deal with the equilibrium point

$$x_0 = \sqrt{b(r-1)},\ y_0 = \sqrt{b(r-1)},\ z_0 = r - 1,$$

which leads to more interesting results. The Jacobian matrix has the characteristic polynomial

$$\lambda^3 + (\sigma + b + 1)\lambda^2 + b(r + \sigma)\lambda + 2b\sigma(r - 1).$$

As polynomial of third degree, the characteristic polynomial has a real root α. On the other hand, the third degree polynomial with the root α and pure imaginary roots $\pm i\beta$, has the expression

$$\lambda^3 - \alpha\lambda^2 + \beta^2\lambda - \alpha\beta^2.$$

By identifying the two polynomials, the critical value r_0 of the parameter r must satisfies the relations

$$\alpha = -(\sigma+b+1),\ \pm\beta = \pm\sqrt{b(r_0+\sigma)},\ 2b\sigma(r_0-1) = b(r_0+\sigma)(\sigma+b+1).$$

From the last relation it follows

$$r_0 = \frac{\sigma(\sigma+b+3)}{\sigma-b-1}$$

and

$$b(r_0+\sigma) = \frac{2b\sigma(\sigma+1)}{\sigma-b-1}.$$

Both numbers are positive if $\sigma > b+1$.

Suppose $\sigma > 0$, $b > 0$ and hence $r > 1$. Taking into account that the solution λ of the characteristic equation

$$\lambda^3 + (\sigma+b+1)\lambda^2 + b(r+\sigma)\lambda + 2b\sigma(r-1) = 0$$

is function of the variable r, taking the derivative in both members, we find

$$\lambda'(r) = \frac{-b(\lambda+2\sigma)}{3\lambda^2 + 2(\sigma+b+1)\lambda + b(r+\sigma)}.$$

It follows

$$(\Re\,\lambda)'(r_0) = \Re\,\lambda'(r_0) = \frac{b(\sigma-b-1)}{2(\beta^2(r_0) + (\sigma+b+1)^2)} > 0.$$

Thus we have shown that we are in the conditions of the Hopf theorem and thus the loss of stability at point r_0 leads to the Hopf-type bifurcation.

3.6 Kepler problem as dynamical system

Let us now consider the problem of the two bodies in space. Suppose that we have two masses located at the points $x_1, x_2 \in \mathbb{R}^3$. They interact with the force F depending only on the distance between the masses and its direction is given by the straight line determined by the positions of the two particles. The kinetic energy is

$$T(\dot{x}) = \frac{m_1}{2}\,\dot{x}_1^2 + \frac{m_2}{2}\,\dot{x}_2^2 = \frac{m_1}{2}\,\langle \dot{x}_1, \dot{x}_1\rangle + \frac{m_2}{2}\,\langle \dot{x}_2, \dot{x}_2\rangle$$

and the potential energy is

$$U(x) = U(\|x_1 - x_2\|).$$

It follows the Lagrangian

$$L(\dot{x}, x) = T(\dot{x}) - U(x),$$

which is invariant with respect to translations (in $\mathbb{R}^6$)

$$(x_1, x_2) \to (x_1 + sa, x_2 + sa),\ a \in \mathbb{R}^3.$$

Since the components of the total moment $m_1\dot{x}_1 + m_2\dot{x}_2$ are first integrals of motion (Euler-Lagrange equations), we can choose new coordinates

$$q_1 = \frac{m_1 x_1 + m_2 x_2}{m_1 + m_2} \text{ (center of mass)},\ q_2 = x_1 - x_2 \text{ (position vector)}$$

in which the Lagrangian is written as

$$L(\dot{q}, q) = \frac{M}{2}\,\dot{q}_1^2 + \frac{\mu}{2}\,\dot{q}_2^2 - U(q_2),\ M = m_1 + m_2,\ \mu = \frac{m_1 m_2}{M}.$$

The Euler-Lagrange system is decoupling

$$\ddot{q}_1 = 0,\ \ddot{q}_2 = -\frac{dU}{dq_2}\,.$$

From the first equation it follows $q_1(t) = q_1(0) + \dot{q}_1(0)t$, i.e., the center of mass has uniform rectilinear motion.

In the sense of scalar product, we multiply the second equation by $2\dot{q}_2$. Integrating we find $\|\dot{q}_2\|^2 = -2U(q_2) + c$. Denoting $r = \|q_2\|$, we deduce

$$\frac{d}{dt}\,r^2 = \frac{d}{dt}\langle q_2, q_2\rangle = 2\langle q_2, \dot{q}_2\rangle.$$

On the other hand,

$$-r\|\dot{q}_2\| = -\|q_2\|\|\dot{q}_2\| \le \langle q_2, \dot{q}_2\rangle \le \|q_2\|\|\dot{q}_2\| = r\|\dot{q}_2\|.$$

We get

$$-\sqrt{-2U(r) + c} \le \frac{dr}{dt} \le \sqrt{-2U(r) + c}.$$

In the case of Kepler problem, the gravitational potential is $U(r) = -\frac{\gamma}{r}$. By integration, it follows two inequalities satisfied by $r(t)$, but we do not recognize the shape of the trajectory.

Another way Having an autonomous Lagrangian, the Hamiltonian (total energy)

$$H(\dot{q}, q) = \frac{M}{2}\,\dot{q}_1^2 + \frac{\mu}{2}\,\dot{q}_2^2 + U(q_2)$$

is a first integral.

Let us formulate physico-mathematical artifacts that lead to the solution of the system of the two differential equations, in a form that shows that the trajectory is a conic located in a special plane. First, the second equation is rotation invariant. It turns out that the angular momentum $I = \mu q_2 \times \dot{q}_2$ is a first integral. We have three first integrals f^1, f^2, f^3, but they are not in involution because (Poisson brackets)

$$\{f^1, f^2\} = f^3,\ \{f^1, f^3\} = -f^2,\ \{f^2, f^3\} = f^1.$$

Using $\{I, ||I||^2\} = 0$, it follows: if $p^1 = M\dot{q}^1$, $p^2 = M\dot{q}^2$ and $p^1 \times p^2$ are the components p^{11}, p^{12}, p^{13}, then

$$p^{11},\, p^{12},\, p^{13},\, \frac{\mu}{2}p^{22} + U(q^2),\, ||I||^2,\, f^3$$

are 6 functional independent first integrals in involution of the two-body problem. Since the motion is made in a plane perpendicular to the moment I, we choose the polar coordinates (r, φ) in this plane. The norm of the angular momentum becomes $I_0 = ||I|| = \mu r^2 \dot{\varphi}$ and the conservation of energy implies the ODE

$$\frac{\mu}{2}\left(\dot{r}^2 + \frac{l_0^2}{\mu^2 r^2}\right) + U(r) = E,$$

which we rewrite into the form

$$\dot{r} = \sqrt{\frac{2(E - U(r))}{\mu} - \frac{l_0^2}{\mu^2 r^2}}\,.$$

In case of Kepler problem the gravitational potential is $U(r) = -\dfrac{\gamma}{r}$. Separating variables, we can calculate the primitive but we cannot find explicitly r as a function of t. That is why we direct our efforts to determine the shape of the orbit, in the form $r = r(\varphi)$, through the differential equation

$$\frac{1}{r^2}\frac{dr}{d\varphi} = \sqrt{\frac{2\mu(E - U(r))}{l_0^2} - \frac{1}{r^2}}\,.$$

The solution is

$$r(\varphi) = \frac{p}{1 - \varepsilon\cos(\varphi - \varphi_0)}, \ p = \frac{l_0^2}{\gamma\mu}, \ \varepsilon = \sqrt{1 + \frac{2El_0^2}{\mu\gamma^2}}.$$

For $\varepsilon < 1$, the orbit is an ellipse; for $\varepsilon = 1$, we find a parabola; and for $\varepsilon > 1$, we have a hyperbola.

3.7 Self-evaluation problems

3.7.1 Stability of equilibrium points for linear differential systems

Problem 20. *Study the stability of the equilibrium points for the linear differential system*

$$\dot{x_1} = -x_1 + ax_2 + bx_3, \ \dot{x_2} = -ax_1 - x_2 + ax_3, \ \dot{x_3} = -bx_1 - ax_2 - x_3.$$

Solution. The equilibrium points are the solutions of the algebraic system

$$-x_1 + ax_2 + bx_3 = 0, -ax_1 - x_2 + ax_3 = 0, -bx_1 - ax_2 - x_3 = 0.$$

The only point of equilibrium is $(0,0,0)$ since the determinant of the algebraic system is non-zero; it satisfies the inequality $\Delta = -1 - b^2 - 2a^2 < 0$.

To decide the stability, we analyse the eigenvalues of the system matrix. The characteristic polynomial is $P(\lambda) = (-1-\lambda)[b^2 + 2a^2 + (-1-\lambda)^2]$, and the eigenvalues are -1 and $-1 \pm i\sqrt{2a^2 + b^2}$. As all have strictly negative real parts, it turns out that $(0,0,0)$ is an asymptotically stable equilibrium point.

Problem 21. *Analyse the stability of the equilibrium points of differential system*

$$\dot{x}_1 = -x_1 + ax_2, \ \dot{x}_2 = bx_1 - x_2 + ax_3, \ \dot{x}_3 = -x_3 + bx_2.$$

Solution. To find the equilibrium points, we solve the algebraic system

$$-x_1 + ax_2 = 0, bx_1 - x_2 + ax_3 = 0, -x_3 + bx_2 = 0.$$

By substitution, we obtain the equation $(2ab - 1)x_2 = 0$. There are two cases:

a) if $ab \neq \frac{1}{2}$, then the dynamical system has a unique equilibrium point, $(0,0,0)$;

b) if $ab = \frac{1}{2}$, then all points of the form $(a\alpha, \alpha, b\alpha)$, with $\alpha \in \mathbb{R}$, are equilibrium points.

The differential system is linear, so to study the stability of the equilibrium points, we find out the eigenvalues of its matrix. The characteristic polynomial is $P(\lambda) = \det(A - \lambda I_3) = (1+\lambda)[2ab - (1+\lambda)^2]$.

- if $ab \neq \frac{1}{2}$ and $ab > 0$, then the eigenvalues are

$$\lambda_1 = -1 < 0,\ \lambda_2 = -1 - \sqrt{2ab} < 0,\ \lambda_3 = -1 + \sqrt{2ab}.$$

  For $ab > \frac{1}{2}$, we have $\lambda_3 > 0$, hence $(0,0,0)$ is an unstable equilibrium point; for $ab < \frac{1}{2}$, it follows $\lambda_i < 0$, $\forall i$, hence $(0,0,0)$ is an asymptotically stable equilibrium point;

- if $ab < 0$, then the eigenvalues are $\lambda_1 = -1 < 0$, $\lambda_{2,3} = -1 \pm i\sqrt{-2ab}$, with $\Re(\lambda_{2,3}) = -1 < 0$, hence $(0,0,0)$ is an asymptotically stable equilibrium point;

- if $ab = 0$, then the only proper value is $\lambda = -1 < 0$, with algebraic multiplicity = geometric multiplicity = 3, hence $(0,0,0)$ is an asymptotically stable equilibrium point;

- if $ab = \frac{1}{2}$, then the eigenvalues are $\lambda_1 = -1 < 0$, $\lambda_2 = -2 < 0$ and $\lambda_3 = 0$, with $m_a = m_g = 1$, whence we deduce that the points $(a\alpha, \alpha, b\alpha)$, with $\alpha \in \mathbb{R}$ are stable equilibrium points, but not asymptotically stable.

3.7.2 Stability of equilibrium points for nonlinear differential systems

Problem 22. *Using the linear approximation, study the stability of the equilibrium point* $(0,0)$ *of the dynamical system* $\dot{x} = e^{x+2y} - \cos 3x$, $\dot{y} = \sqrt{4+8x} - 2e^{y}$.

Solution. The equilibrium points are found solving the algebraic system

$$e^{x+2y} - \cos 3x = 0, \sqrt{4+8x} - 2e^y = 0.$$

We observe that $(0,0)$ is the only equilibrium point.

Denote $f_1(x,y) = e^{x+2y} - \cos 3x,\ f_2(x,y) = \sqrt{4+8x} - 2e^y$. The matrix of the linearized system around the equilibrium point is

$$A = \begin{pmatrix} \frac{\partial f_1}{\partial x}(0,0) & \frac{\partial f_1}{\partial y}(0,0) \\ \frac{\partial f_2}{\partial x}(0,0) & \frac{\partial f_2}{\partial y}(0,0) \end{pmatrix} = \begin{pmatrix} e^{x+2y} + 3\sin 3x & 2e^{x+2y} \\ \frac{2}{\sqrt{1+2x}} & -2e^y \end{pmatrix}_{(0,0)},$$

i.e., $A = \begin{pmatrix} 1 & 2 \\ 2 & -2 \end{pmatrix}$. The characteristic polynomial is $\det(A - \lambda I_3) = \lambda^2 + \lambda - 6$, with real eigenvalues 2 and -3. Since $\lambda_1 > 0$, it follows that $(0,0)$ is an unstable equilibrium point.

Problem 23. *Using the linear approximation, study the stability of the equilibrium point for the dynamical system* $\dot{x} = 2e^{-x} - \sqrt{4+ay}$, $\dot{y} = \ln(1+9x+ay)$.

Solution. The equilibrium point is the solution of the algebraic system

$$2e^{-x} - \sqrt{4+ay} = 0,\ \ln(1+9x+ay) = 0.$$

Obviously, $(0,0)$ is an equilibrium point. We can prove that it is unique in this way: by substitution method, we attend the equation $2e^{-x} = \sqrt{4-9x}$ and it is shown, using classical techniques from mathematical analysis, the fact that the last equation has the unique solution $x = 0$.

The matrix of linearized system is

$$A = \begin{pmatrix} -2e^{-x} & -\frac{a}{2\sqrt{4+ay}} \\ \frac{9}{1+9x+ay} & \frac{a}{1+9x+ay} \end{pmatrix}_{(0,0)} = \begin{pmatrix} -2 & -\frac{a}{4} \\ 9 & a \end{pmatrix}.$$

The characteristic polynomial is $P(\lambda) = \lambda^2 + (2-a)\lambda + \frac{a}{4}$, with $\Delta = (a-1)(a-4)$. A discussion is required according to the parameter a:

- if $a \in (-\infty, 1) \cup (4, \infty)$, then the eigenvalues are real numbers, $\lambda_{1,2} = \dfrac{a - 2 \pm \sqrt{a^2 - 5a + 4}}{2}$ and their sign is studied in turn according to parameter a.

  - for $a \in (-\infty, 0) \cup (4, \infty)$, we have $\lambda_1 > 0$, hence $(0, 0)$ is an unstable equilibrium point for the linearized system and for the initial, non-linear system;

  - for $a = 0$, we find $\lambda_1 = 0$, $\lambda_2 < 0$, hence the point $(0, 0)$ is not a hyperbolic point and its stability cannot be decided using linear approximation (since $\lambda_1 = 0$ has $m_a = m_g = 1$, the equilibrium point $(0, 0)$ of the linearized system is stable, but not asymptotically stable);

  - for $a \in (0, 1)$, we have $\lambda_{1,2} < 0$, hence $(0, 0)$ is an asymptotically stable equilibrium point both for the linearized system and for the non-linear system;

- if $a = 1$, then $\lambda_{1,2} = -\dfrac{1}{2} < 0$, hence $(0, 0)$ is an asymptotically stable equilibrium point both for the linearized system and for the non-linear system;

- if $a = 4$, then $\lambda_{1,2} = 1 > 0$, hence $(0, 0)$ is an unstable equilibrium point for both systems;

- if $a \in (1, 4)$, then the eigenvalues are complex conjugate,

$$\lambda_{1,2} = \frac{a - 2 \pm i\sqrt{-a^2 + 5a - 4}}{2},$$

  with real part $\Re(\lambda_{1,2}) = \dfrac{a-2}{2}$, whose sign varies according to the parameter a, as follows:

  - for $a \in (1, 2)$, we have $\Re(\lambda_{1,2}) < 0$, hence $(0, 0)$ is an asymptotically stable equilibrium point for both systems;

  - for $a = 2$, we have $\Re(\lambda_{1,2}) = 0$, hence $(0, 0)$ is not an hyperbolic point and its stability for the nonlinear system using linearization cannot be deduced;

  - for $a \in (2, 4)$, we have $\Re(\lambda_{1,2}) > 0$, hence $(0, 0)$ is an unstable equilibrium point for both systems.

In conclusion, the point $(0,0)$ can be: equilibrium point unstable, for $a \in (-\infty, 0) \cup (2, \infty)$; point of stable asymptotic equilibrium, for $a \in (0,2)$. For $a \in \{0; 2\}$, the point $(0,0)$ is not hyperbolic (for the initial, nonlinear system) and its stability cannot be studied using linear approximation.

Problem 24. *Analyse the equilibrium point of the Van der Pol differential equation*

$$\ddot{x} - \mu\,(1 - x^2)\,\dot{x} + x = 0, \quad \mu > 0.$$

Solution. It is a differential equation of second order. Introducing a new unknown function $y = \dot{x}$, we transform it into a system of differential equations of order one,

$$\dot{x} = y, \;\; \dot{y} = \mu\,(1 - x^2)\,y - x.$$

The equilibrium points are the solutions of the algebraic system

$$y = 0, \;\; \mu\,(1 - x^2)\,y - x = 0.$$

Obviously, $(0,0)$ is the only equilibrium point. Since the system is not linear, for the stability study we will resort to its linearization around the point $(0,0)$ and the stability study for the linearized system. Denote $f_1(x,y) = y, \; f_2(x,y) = \mu\,(1-x^2)\,y - x$. The matrix of linearized system is

$$A = \begin{pmatrix} \dfrac{\partial f_1}{\partial x}(0,0) & \dfrac{\partial f_1}{\partial y}(0,0) \\ \dfrac{\partial f_2}{\partial x}(0,0) & \dfrac{\partial f_2}{\partial y}(0,0) \end{pmatrix} = \begin{pmatrix} 0 & 1 \\ -1 & \mu \end{pmatrix}.$$

The characteristic polynomial is $P(\lambda) = \lambda^2 - \mu\lambda + 1$. If $\mu \in (0,2)$, then the eigenvalues $\lambda_{1,2} = \dfrac{\mu \pm i\sqrt{4-\mu^2}}{2}$ have the real part $\Re(\lambda_{1,2}) = \dfrac{\mu}{2} > 0$, hence $(0,0)$ is an unstable equilibrium point. If $\mu \in [2, \infty)$, then $\lambda_{1,2} = \dfrac{\mu \pm \sqrt{\mu^2 - 4}}{2} > 0$, which also implies unstable equilibrium.

Hence $(x = 0, \dot{x} = 0)$ is an unstable equilibrium point for the Van der Pol equation, for any $\mu > 0$.

Problem 25. *The scalar field $f(x, y) = x^3 + y^3 - 3xy$ is given on $\mathbb{R}^2$. Write the associated potential dynamical system and analyse the stability of its equilibrium points.*

Solution. For any scalar field $f : D \subset \mathbb{R}^n \to \mathbb{R}$, the differential system $\frac{dx}{dt} = \text{grad}f$ is called *either potential system, or gradient system* associated to the scalar field f. Its solutions are called *gradient lines* of f. The critical point of f are equilibrium points of the potential system.

We have $\text{grad}f = (3x^2 - 3y, 3y^2 - 3x)$, and hence the associated potential system is

$$\dot{x} = 3x^2 - 3y,\ \dot{y} = 3y^2 - 3x.$$

The equilibrium points are the points that cancel $\text{grad}f$, i.e., $x^2 - y = 0, y^2 - x = 0$. By substitution we obtain the equation $x^4 = x$, with solutions $x = 0$, $x = 1$, whence the equilibrium points $(0, 0)$ and $(1, 1)$.

For the stability study, we use the linearized system. The Jacobi matrix is $A = \begin{pmatrix} 6x & -3 \\ -3 & 6y \end{pmatrix}$.

At $(0, 0)$, the matrix becomes $A = \begin{pmatrix} 0 & -3 \\ -3 & 0 \end{pmatrix}$, with eigenvalues $\lambda_{1,2} = \pm 3$. As one of them is greater than 0, it follows that $(0, 0)$ is an unstable equilibrium point.

At $(1, 1)$, the matrix eigenvalues are 3 and 9, strictly positive, consequently $(1, 1)$ is an unstable equilibrium point.

3.7.3 Lyapunov functions

Problem 26. *Analyse the stability of the equilibrium point $(0, 0)$ of the differential system associated to the vector field $X(x, y) = (2y^3 - x^5, -x - y^3 + y^5)$, using a Lyapunov function.*

Solution. The differential system is

$$\dot{x} = 2y^3 - x^5,\ \dot{y} = -x - y^3 + y^5.$$

The point $(0, 0)$ cancels the components of the field, so it is an equilibrium point. We guess that the function $L : \mathbb{R}^2 \to \mathbb{R},\ L(x, y) = x^2 + y^4$ is what we need. We show that it fulfills the conditions of one of the Lyapunov stability theorems.

We have $L(0,0)=0;\quad L(x,y)>0,\ \forall (x,y)\neq(0,0)$, hence the function L is positive definite. We compute the derivative of the function L along the field lines of the vector field X:

$$D_X L \stackrel{\text{def}}{=} \langle \text{grad} L(x,y), X(x,y)\rangle$$

$$= \langle (2x, 4y^3), (2y^3 - x^5, -x - y^3 + y^5)\rangle = -2(x^6 + 2y^6 - 2y^8).$$

We are looking for a neighbourhood U of the equilibrium point $(0,0)$ on which $D_X L$ is negative definite or semi-definite. For example, we want $y^6 \geq y^8$ on an open set around the point 0. This inequality is true for $y \in (-1,1)$. Then we have $D_X L(0,0)=0$ and $D_X L(x,y) < 0,\ \forall (x,y)\in U\setminus\{(0,0)\}$, hence $D_X L$ is negative definite on the set U.

Therefore, we restrict the function L to the neighbourhood $U = \mathbb{R}\times(-1,1)$ of $(0,0)$. Because L is positive definite, and $D_X L$ is negative definite, $(0,0)$ is an asymptotically stable equilibrium point.

Problem 27. *Find a Lyapunov function and study stability of the equilibrium point* $(0,0)$ *for the dynamical system associated to the vector field*

$$X(x,y) = \left(-y - \frac{x}{2} - \frac{x^3}{4},\ x - \frac{y}{2} - \frac{y^3}{4}\right).$$

Solution. The differential system generated by the given vector field is

$$\dot{x} = -y - \frac{x}{2} - \frac{x^3}{4},\ \dot{y} = x - \frac{y}{2} - \frac{y^3}{4}.$$

The equilibrium point of the system is $(0,0)$. We guess that the function $L:\mathbb{R}^2\to\mathbb{R},\ L(x,y)=x^2+y^2$ is a candidate for what we need. We show that it fulfills the conditions of one of the Lyapunov stability theorems.

We have $L(0,0)=0;\quad L(x,y)>0,\ \forall(x,y)\neq(0,0)$, hence the function L is positive definite. The derivative of the function L along the field lines of X is

$$D_X L \stackrel{\text{def}}{=} \langle \text{grad} L(x,y), X(x,y)\rangle$$

$$= \langle (2x,2y), \left(-y - \frac{x}{2} - \frac{x^3}{4}, x - \frac{y}{2} - \frac{y^3}{4}\right)\rangle = -\left(x^2 + \frac{x^4}{2} + y^2 + \frac{y^4}{2}\right).$$

Then we have $D_X L(0,0) = 0$ and $D_X L(x,y) < 0,\ \forall (x,y)\in\mathbb{R}^2\setminus\{(0,0)\}$, hence $D_X L$ is negative definite on $\mathbb{R}^2$.

Since L is positive definite, and $D_X L$ is negative definite, the point $(0,0)$ is an asymptotically stable equilibrium point.

Problem 28. *Analyse the stability of the equilibrium point of the differential system that correspond to the vector field* $X(x, y) = (x^2 + y^2, -2xy)$*, using a Lyapunov function.*

Solution. The differential system determined by the given vector field is

$$\dot{x} = x^2 + y^2, \ \dot{y} = -2xy.$$

Obviously, $(0, 0)$ is the only equilibrium point of the system. We try the function $L : \mathbb{R}^2 \to \mathbb{R}$, $L(x, y) = -x^3 + 3xy^2$. We have to check the applicability of one of the Lyapunov stability theorems.

We have $L(0, 0) = 0$. To study the sign of the expression $L(x, y)$, we use its decomposition into factors, $L(x, y) = x(y\sqrt{3} - x)(y\sqrt{3} + x)$. The equations $x = 0$, $y\sqrt{3} - x = 0$ and $y\sqrt{3} + x = 0$ designate three straight lines concurrent at $(0, 0)$. They divide the plane into 6 regions where the sign of the function L alternates, regions that have the point $(0, 0)$ on the boundary. Thus, in no neighbourhood of the point $(0, 0)$, the function cannot have a constant sign, because any neighbourhood contains points from all the regions above. So there is no neighborhood of $(0, 0)$ on which L is positive definite.

We shall apply the following result:

Let X be a vector field and x_0 its equilibrium point. We consider the function $L : U \subseteq \mathbb{R}^n \to \mathbb{R}$, of class C^1, for which $L(x_0) = 0$, where U is a neighborhood of the point x_0. If there exists U^+, an open subset in U such that $L(x) > 0$ and $D_X L(x) > 0, \ \forall x \in U^+$, and $x_0 \in \partial U^+$ (i.e., x_0 belong to the boundary of the set U^+), then x_0 is an unstable equilibrium point.

One calculates the derivative of the function L along field lines of X,

$$D_X L \stackrel{\text{def}}{=} \langle \text{grad} L(x, y), X(x, y) \rangle$$

$$= \langle (-3x^2 + 3y^2, 6xy), (x^2 + y^2, -2xy) \rangle = 3(y^4 - 4x^2y^2 - x^4).$$

We are looking for an open set U^+ of $\mathbb{R}^2$, whose boundary contains the equilibrium point $(0, 0)$ and on which L and $D_X L$ to be positive definite. For that, we need a decomposition in factors of the derivative

$$D_X L = 3(y^4 - 4x^2y^2 + 4x^4 - 5x^4) = 3[(y^2 - 2x^2)^2 - (\sqrt{5}x^2)^2]$$

$$= 3\left(y - \sqrt{2 + \sqrt{5}}\, x\right)\left(y + \sqrt{2 + \sqrt{5}}\, x\right)\left(y^2 + (\sqrt{5} - 2)^2\right).$$

The last factor is positive, and the other two define, as factors in decomposition of $D_X L$, some straight lines that pass through the equilibrium point. We are looking for an open set U^+ included in the first quadrant, such that the factors $x(y\sqrt{3}+x)$ and $y+\sqrt{2+\sqrt{5}}\,x$ will be immediately positive. It rests to find an open set in the first quadrant, with the point $(0,0)$ on boundary, on which we have $y\sqrt{3}-x>0$ and $y-\sqrt{2+\sqrt{5}}\,x>0$. Since $\sqrt{2+\sqrt{5}}>\dfrac{1}{\sqrt{3}}$, it follows that the second condition ensures fulfillment of the first condition. We consider

$$U^+ = \{(x,y)\in\mathbb{R}^2 \mid y > x\sqrt{2+\sqrt{5}},\ x>0\},$$

looked up open set, with the equilibrium point $(0,0)$ on boundary, such that L and $D_X L$ are positive definite on U^+. Then $(0,0)$ is an unstable equilibrium point.

Problem 29. *Analyse the stability of the equilibrium points of Lorenz differential system generated by the vector field*

$$X(x,y,z) = (\sigma(y-x),\ rx-y-xz,\ xy-bz),\ \ \sigma,r,b>0,$$

using the Lyapunov function $L(x,y,z)=rx^2+\sigma y^2+\sigma z^2$.

Solution. To find the equilibrium points, we solve the algebraic system

$$\sigma(y-x)=0, rx-y-xz=0, xy-bz=0.$$

From the first equation it follows $x=y$. Going to the second equation and then to the third, on the one hand we find the equilibrium point $(0,0,0)$, on the other hand we obtain the relations $z=r-1$ and $x^2=b(r-1)$, where two cases are distinguished:

a) If $r\leq 1$, then the only equilibrium point is $(0,0,0)$ (with the remark that for $r=1$, the equation $x^2=b(r-1)$ has also the solution $(0,0,0)$).

For the study of stability, we shall use the Lyapunov function $L(x,y,z)=rx^2+\sigma y^2+\sigma z^2>0,\ \forall(x,y,z)\neq(0,0,0)$; Furthermore, $L(0,0,0)=0$. We compute the derivative along the field lines of X. We have

$$D_X L = \langle \mathrm{grad} L, X\rangle = \langle (2rx, 2\sigma y, 2\sigma z), (\sigma(y-x), rx-y-xz, xy-bz)\rangle$$

$$= -2\sigma[z^2+r(x-y)^2+(1-r)y^2]\leq 0,\ \forall(x,y,z)\in\mathbb{R}^3-\{(0,0,0)\},$$

(since $\sigma, r, b > 0$ and $r \leq 1$), whence we deduce that the point $(0,0,0)$ is a stable equilibrium point. In the case $r < 1$, the inequality is strictly on $\mathbb{R}^3 \setminus \{(0,0,0)\}$, hence $(0,0,0)$ is an asymptotically stable equilibrium point.

b) If $r > 1$, then $x = y = \pm\sqrt{b(r-1)}$, which leads to two new equilibrium points, $(\pm\sqrt{b(r-1)}, \pm\sqrt{b(r-1)}, r-1)$. For the stability study, we use the linearization around $(0,0,0)$, which has the matrix $A = \begin{pmatrix} -\sigma & \sigma & 0 \\ r & -1 & 0 \\ 0 & 0 & -b \end{pmatrix}$, with eigenvalues $\lambda_1 = -b < 0$,

$$\lambda_2 = \frac{-\sigma - 1 - \sqrt{(\sigma-1)^2 + 4\sigma r}}{2} < 0,$$

$$\lambda_3 = \frac{-\sigma - 1 + \sqrt{(\sigma-1)^2 + 4\sigma r}}{2} > 0$$

and as one of them is strictly positive, it turns out that in the case $r > 1$, the equilibrium point $(0,0,0)$ is unstable.

The matrix of the linearized system around the equilibrium points $(\pm\sqrt{b(r-1)}, \pm\sqrt{b(r-1)}, r-1)$ is

$$A = \begin{pmatrix} -\sigma & \sigma & 0 \\ 1 & -1 & \mp\sqrt{b(r-1)} \\ \pm\sqrt{b(r-1)} & \pm\sqrt{b(r-1)} & -b \end{pmatrix},$$

with the characteristic polynomial $-\lambda^3 - (b+\sigma+1)\lambda^2 - b(r+\sigma)\lambda - 2\sigma b(r-1)$, being difficult to study its roots. After laborious calculations, it can be shown that for $r < \frac{47}{19}$, the two points are asymptotically stable equilibrium points, all the eigenvalues having the real part strictly negative.

Problem 30. *(Continuation) Classify, from the point of view of stability, the equilibrium points of the Lorenz dynamical system, generated by the vector field*

$$X(x) = (x_2 - x_1,\ rx_1 - x_2 - x_1x_3,\ x_1x_2 - x_3),\ \ r > 0.$$

Solution. The problem is a particularization of the previous problem, in which $\sigma = b = 1$ was considered. We have shown that for $r \leq 1$, the only equilibrium point is $(0,0,0)$. It is a stable equilibrium point (for $r < 1$, even asymptotically stable).

For $r > 1$, the equilibrium point $(0,0,0)$ is unstable, and the study of the stability of the two equilibrium points becomes accessible.

The equilibrium points arc $(\pm\sqrt{r-1}, \pm\sqrt{r-1}, r-1)$. We use the linearization. The matrix of the linearized system around these points has the charateristic polynomial $P(\lambda) = -\lambda^3 - 3\lambda^2 - (r+1)\lambda - 2(r-1)$. We remark that $\lambda_1 = -2 < 0$ is an eigenvalue. The others eigenvalues are roots of the polynomial $-\lambda^2 - \lambda + (1-r)$. It follows that $\lambda_2 + \lambda_3 = -1 < 0$ and $\lambda_2\lambda_3 = r - 1 > 0$. Hence $\Re\lambda_{2,3} < 0$, i.e., the equilibrium points are asymptotically stable.

Thus, the remaining points not studied in the previous problem, in the case of $r > 1$, are, for the particular choice $\sigma = b = 1$, asymptotically stable equilibrium points.

Problem 31. *Study the stability of a general damped harmonic oscillator that is described by*

$$m\,\ddot{y}(t) + f(\dot{y}(t)) + k\,y(t) = 0,$$

using a Lyapunov function.

Solution. The dumping $f(\dot{y}(t))$ is a nonlinear function of velocity. The total energy of such a system is

$$E(y(t), \dot{y}(t)) = \frac{1}{2}\,k\,y^2(t) + \frac{1}{2}\,m\,\dot{y}^2(t).$$

We find

$$\frac{d}{dt}E(y(t), \dot{y}(t)) = -f(\dot{y}(t))\dot{y}(t).$$

The condition $f(\dot{y}(t))\dot{y}(t) > 0$ (the dumping term is that it opposes the motion, so that the graph of f should lie in the first and third quadrants) would indicate that the harmonic oscillator would be stable.

Problem 32. *For the differential system associated to the vector field*

$$X(x,y) = \left(y - ax\left(x^2 + y^2\right), -x - ay\left(x^2 + y^2\right)\right),$$

study the stability of the equilibrium point, using a Lyapunov function.

Solution. The equilibrium points are the solutions of the algebraic system

$$y - ax(x^2 + y^2) = 0, \;\; -x - ay(x^2 + y^2 = 0.$$

The only equilibrium point is $(0,0)$.

Let us try to use $L(x,y) = x^2 + y^2$ as a Lyapunov function. Obviously, L is positive definite. Its derivative along the direction of the vector field X is

$$D_X L = \langle \text{grad} L(x,y), X(x,y) \rangle = \langle (2x, 2y), (y - ax(x^2+y^2), -x - ay(x^2+y^2)) \rangle$$

$$= -2a(x^2 + y^2)^2.$$

A discussion is required according to the parameter a:

a) if $a > 0$, then $D_X L < 0$, $\forall (x,y) \in \mathbb{R}^2 - \{(0,0)\}$, whence it follows that $(0,0)$ is an asymptotically stable equilibrium point;

b) if $a < 0$, then $D_X L > 0$, $\forall (x,y) \in \mathbb{R}^2 - \{(0,0)\}$, hence $(0,0)$ is an unstable equilibrium point;

c) if $a = 0$, then the system becomes $\dot{x} = y$, $\dot{y} = -x$, which is a linear system. The eigenvalues of the matrix of the system are $\pm i$, with real part 0, but because they have algebraic multiplicity equal to the geometric one (equal to 1), it turns out that $(0,0)$ it is a stable equilibrium point, but not asymptotically stable.

Problem 33. *We consider the dynamical system*

$$\dot{x} = x - y - x(x^2+y^2) + \frac{xy}{\sqrt{x^2+y^2}}, \quad \dot{y} = x + y - y(x^2+y^2) - \frac{x^2}{\sqrt{x^2+y^2}}.$$

Find the equilibrium points and analyse their stability.

Solution. The equilibrium points are solutions of the algebraic system

$$x - y - x(x^2+y^2) + \frac{xy}{\sqrt{x^2+y^2}} = 0, \quad x + y - y(x^2+y^2) - \frac{x^2}{\sqrt{x^2+y^2}} = 0.$$

If $x = 0$, from the first equation it follows $y = 0$, but the condition $x^2 + y^2 \neq 0$ is required.

If $y = 0$, the algebraic system becomes

$$x - x^3 = 0, \quad x - \frac{x^2}{|x|} = 0,$$

with unique solution $x = 1$. Hence $(1,0)$ is an equilibrium point.

Now we consider $x \neq 0$, $y \neq 0$. We multiply the first equation by x, and the second by y. Adding, we obtain

$$(x^2 + y^2)(1 - x^2 - y^2) = 0,$$

whence it follows $x^2+y^2=1$. Returning to the first equation, we find $x=1$, which leads to $y=0$, contradiction with the case where we are.

So $(1,0)$ is the only equilibrium point. We study stability. Linearization cannot be used, because the point is not hyperbolic (one of the eigenvalues of the matrix of the linearized system is zero). We are looking for a Lyapunov function, but the expressions are complicated, and the Lyapunov function is difficult to guess. To do this, we make a change of unknown functions, moving to polar coordinates, so that the system becomes simpler. We have

$$x(t)=\rho(t)\cos\theta(t),\ \ y(t)=\rho(t)\sin\theta(t),$$

with $\rho>0$ (since $(x,y)\neq(0,0)$) and $\theta\in[0,2\pi)$. To the point $(1,0)$ in Cartesian coordinates corresponds the point $(1,0)$ in polar coordinates. With the substitutions above, the system becomes

$$\begin{cases}\dot\rho\cos\theta-\rho\dot\theta\sin\theta=\rho\cos\theta-\rho\sin\theta-\rho^3\cos\theta+\dfrac{\rho^2\sin\theta\cos\theta}{\rho}\\[2mm] \dot\rho\sin\theta+\rho\dot\theta\cos\theta=\rho\cos\theta+\rho\sin\theta-\rho^3\sin\theta-\dfrac{\rho^2\cos^2\theta}{\rho}.\end{cases}$$

If we multiply the first equation by $-\sin\theta$ and the second by $\cos\theta$, so adding we find the differential equation $\dot\theta=1-\cos\theta$. Multiplying the first equation by $\cos\theta$ and the second by $\sin\theta$, we obtain the differential equation $\dot\rho=\rho-\rho^3$.

Thus the system expressed in polar coordinates is

$$\dot\rho=\rho-\rho^3,\ \ \dot\theta=1-\cos\theta,$$

with equilibrium point $(1,0)$.

We select the function $L(\rho,\theta)=\theta-\sin\theta$. We verify the fulfillment of the conditions of one of the Lyapunov stability theorems.

We have $L(1,0)=0$. We calculate the derivative of this function following the direction of the vector field $X(\rho,\theta)=(\rho-\rho^3,1-\cos\theta)$. We find

$$\begin{aligned}D_XL&=\langle\mathrm{grad}L(\rho,\theta),X(\rho,\theta)\rangle=\langle(0,1-\cos\theta),(\rho-\rho^3,1-\cos\theta)\rangle\\&=(1-\cos\theta)^2>0\end{aligned}$$

on the open set $(0,1)\times(0,2\pi)$.

On the open set $U^+=(0,1)\times(0,2\pi)$, the inequalities $L(\rho,\theta)=\theta-\sin\theta>0$, $D_XL>0$, are true and the point $(1,0)$ belongs to the boundary of U^+. It follows that $(1,0)$ is an unstable equilibrium point.

Problem 34. *For the function $f(x,y) = xy(a-x-y)$, write the associated potential dynamical system and discuss the stability of its equilibrium points.*

Solution. Since $\text{grad} f = (y(a-2x-y), x(a-x-2y))$, the potential system is

$$\dot{x} = y(a-2x-y), \ \dot{y} = x(a-x-2y).$$

The equilibrium points are solutions of the system $y(a-2x-y) = 0, x(a-x-2y) = 0$. It follows $(0,0)$, $(a,0)$, $(0,a)$, $\left(\frac{a}{3}, \frac{a}{3}\right)$, that coincides for $a = 0$.

Case $a \neq 0$: the potential system has 4 distinct equilibrium points, and stability is studied one by one, using linearization.

The matrix of partial derivatives is

$$A = \begin{pmatrix} -2y & a-2x-y \\ a-2x-y & -2x \end{pmatrix},$$

which we calculate in turn of each of the above points.

i) For the point $(0,0)$, the matrix is $A = \begin{pmatrix} 0 & a \\ a & 0 \end{pmatrix}$, with eigenvalues $\lambda_{1,2} = \pm a$, of which at least one is strictly positive, so $(0,0)$ is an unstable equilibrium point for the linear system, so for the initial system.

ii) For the point $(a,0)$, the matrix is $A = \begin{pmatrix} 0 & -a \\ -a & -2a \end{pmatrix}$, with real eigenvalues (we have $\Delta = 8a^2 > 0, \ a \neq 0$) and of opposite signs (we have $\lambda_1\lambda_2 = -a^2 < 0$), so one is strictly positive. It turns out that $(a,0)$ is an unstable equilibrium point for the linear system, so for the nonlinear system.

iii) Analogously for the point $(0,a)$.

iv) For the point $\left(\frac{a}{3}, \frac{a}{3}\right)$, the matrix is $A = \begin{pmatrix} -\frac{2a}{3} & -\frac{a}{3} \\ -\frac{a}{3} & -\frac{2a}{3} \end{pmatrix}$,

with distinct real eigenvalues $\lambda_1 = -\frac{a}{3}$, $\lambda_2 = -a$. If $a > 0$, then the eigenvalues are strictly negative, whence we find that $\left(\frac{a}{3}, \frac{a}{3}\right)$ is an asymptotically stable point both for linearized system and the initial nonlinear potential system. If $a < 0$, then the eigenvalues are strictly positive, whence we deduce that $\left(\frac{a}{3}, \frac{a}{3}\right)$ is an unstable equilibrium point both for the linearized system and for the nonlinear potential system.

Case $a = 0$**:** then the only equilibrium point of the potential system is $(0,0)$. To study the stability, we cannot use linearization, because $(0,0)$ is not a hyperbolic point, the matrix having null eigenvalues.

We are looking for a Lyapunov function, say $L(x,y) = x^2 + y^2$. We have $L(0,0) = 0$, $L(x,y) > 0$, $\forall (x,y) \neq (0,0)$.

We compute the derivative along field lines of $\text{grad} f = (-2xy - y^2, -x^2 - 2xy)$. This derivative,

$$D_{\text{grad}f} L = \langle \text{grad} L, \text{grad} f \rangle = \langle (2x, 2y), (-2xy - y^2, -x^2 - 2xy) \rangle$$

$$= -6xy(x+y),$$

cannot have constant sign on any neighbourhood of $(0,0)$, since the factors have variable signs around zero. But on the third quadrant, for example, we have $x < 0$, $y < 0$, $x + y < 0$, hence $-6xy(x+y) > 0$. Consequently, we consider the open set $U^+ = \{(x,y) \in \mathbb{R}^2 | x < 0,\ y < 0\}$. On U^+, we have $L > 0$, $D_{\text{grad}f} L > 0$, and $(0,0) \in \partial U^+$. It follows that $(0,0)$ is an unstable equilibrium point.

3.8 Research problems

Problem 35. *Show that the solution of the Cauchy Riccati matrix problem*

$$\dot{H} = AH + HA - 2HAH,\ H(0) = H_0$$

is

$$H(t) = e^{tA} H_0 (I_n - H_0 + e^{2tA} H_0)^{-1} e^{tA},\ t \in \mathbb{R}.$$

Find the equilibrium positions and investigate their stability.

Problem 36. *A dynamical system of the form*

$$\dot{x}(t) = V_0(x(t)) + \varepsilon V_1(x(t), \varepsilon),\ x \in \mathbb{R}^n,\ \varepsilon \in [0, \delta)$$

is called disturbance system of $\dot{x}(t) = V_0(x(t))$, $x \in \mathbb{R}^n$, *obtained for* $\varepsilon = 0$. *What form has the disturbance system in polar coordinates,* $x \to I, \varphi$; $I = (I^1, ..., I^{n-m}) \in \mathbb{R}^{n-m}, \varphi = (\varphi^1, ..., \varphi^m) \in T^m$ *(where* T^m *is the m-dimensional torus,* I *is the slow variable and* φ *is the fast variable)?*

Hint $\dot{I} = \varepsilon f(I, \varphi, \varepsilon)$, $\dot{\varphi} = \omega(I) + \varepsilon g(I, \varphi, \varepsilon)$.

Problem 37. *(Continuation, mediation principle) The time behaviour of the slow variables is obtained by replacing the first equation of the perturbed system, in polar coordinates, with the mediated equation*

$$\dot{J} = \varepsilon F(J),\ F(J) = \frac{1}{(2\pi)^m} \int_0^{2\pi} \dots \int_0^{2\pi} f(I, \varphi, 0) d\varphi^1 \dots d\varphi^m.$$

Compute $F(J)$ in case $f(I, \varphi, \varepsilon) = I \sin\varphi^1 \cos\varphi^2$.

Problem 38. *We consider the van der Pol equation*

$$\ddot{x}(t) = -x(t) + \varepsilon\left(1 - x^2(t)\right)\dot{x}(t),$$

which describes oscillations with small nonlinear friction (negative for large amplitude and positive for small amplitude). (i) Write the unperturbed equation. (ii) Transform the equation into a first order system, analyse the stability of the equilibrium positions. (iii) Pass the coordinate system into polar coordinates, $x = \sqrt{2I}\sin\varphi$, $\dot{x} = \sqrt{2I}\cos\varphi$, and revisit the stability of the equilibrium positions.

Problem 39. *Analysing a circuit using Kirchhoff's circuit laws, the dynamics of Chua's circuit can be accurately modelled by means of a system of three nonlinear ODEs in the unknowns $x(t), y(t), z(t)$, which represent the voltages across the capacitors C_1, C_2 and the electric current in the inductor L_1 respectively. These equations are:*

$$\dot{x}(t) = \alpha[y(t) - x(t) - f(x(t))],\ RC_2\dot{y}(t) = x(t) - y(t) + Rz(t),\ \dot{z}(t) = -\beta y(t).$$

The function $f(x)$ describes the electrical response of the nonlinear resistor, and its shape depends on the particular configuration of its components. The parameters α and β are determined by the particular values of the circuit components. Find a computer-assisted proof of chaotic behaviour (more precisely, of positive topological entropy) in Chua's circuit (see [1]).

Hint We use a three-segment piecewise-linear characteristic function $f(x) = bx + 0.5(a - b)(|x + 1| - |x - 1|)$.

Bibliography

[1] Z. Galias, *Positive topological entropy of Chua's circuit: a computer-assisted proof*, Int. J. Bifurcations and Chaos, 7 (1997), pp. 331-349.

[2] R. Munnoz-Alicea, *Introduction to Bifurcations and The Hopf Bifurcation Theorem for Planar Systems*, www.math.colostate.edu

[3] C. Rogers, W. K. Schief, *Backlund and Darboux Transformations, Geometry and Modern Applications in Soliton Theory*, Cambridge University Press, 2002.

[4] N. Sheridan, *Hamilton's Ricci Flow*, Honours Thesis, The University of Melbourne, Department of Mathematics and Statistics, November 2006.

[5] G. Teschl, *Ordinary Differential Equations and Dynamical Systems*, Viena University, Austria, 2007.

[6] C. Udriste, *Geometric Dynamics*, Mathematics and Its Applications, 513, Kluwer Academic Publishers, Dordrecht, Boston, London, 2000.

[7] C. Udriste, L. Matei, *Lagrange-Hamilton Theories* (in Romanian), Monographs and Textbooks 8, Geometry Balkan Press, Bucharest, 2008.

[8] E. T. Whittaker, *A Treatise on The Analytical Dynamics of Particles & Rigid Bodies*, Cambridge University Press, 1989.

Chapter 4

GYROSCOPIC FORCES AND COLLISION AVOIDANCE

Motto:
Some people hold dear,
What's vain in the other men's eye;
But He who can scan both the earth and the sky
And set up a bridge between life and death,
Will always distinguish
"Black" from "White" in his gifts.

George Coşbuc - *The Poet*

This Chapter introduces and studies gyroscopic forces in connection with collision theory and bypassing convex obstacles by vehicles. Among the concepts used are: detection sensor, safety sensor, transforming a special differential equation into an integral equation, Lyapunov functions, and optimal control. Suggestions for further reading: [1]-[15].

4.1 Gyroscopic forces

The forces $F(x,\dot{x}) = (F_i(x,\dot{x})), i = 1,...,n$, dependent on the point x and the velocity $\dot{x}$, i.e., defined on $\mathbb{R}^{2n}$, which produce zero mechanical work, i.e.,

$$\int_\Gamma F_i(x,\dot{x})\, dx^i = \int_a^b F_i(x(t),\dot{x}(t))\, \dot{x}^i(t)dt = 0,$$

are called *gyroscopic forces.* The simplest example of *gyroscopic force* $F(x,\dot{x})$ is one that satisfies $\langle F(x,\dot{x}),\dot{x}\rangle = F_i(x,\dot{x})\,\dot{x}^i = 0$ (the elementary mechanical work is zero), where $\langle\, ,\, \rangle$ means the scalar product. A fairly general class of gyroscopic forces is of the matrix form

$$F_g(x,\dot{x}) = \Omega(x,\dot{x})\dot{x},$$

where $\Omega(x,\dot{x})$ is an antisymmetric matrix. Coriolis forces (in mechanics) and Lorentz forces (in electromagnetism) are gyroscopic forces.

Generally, gyroscopic forces are useful for stabilizing mechanical systems. In the dynamics of mechanical systems there are two essential points of view on this type of forces. One is that gyroscopic forces create couples between different degrees of freedom, just as in mechanical couples. Another is that gyroscopic forces rotate the velocity vector just as the magnetic field rotates the velocity of a charged particle. Firstly, the matrix $\Omega(x,\dot{x})$ can be interpreted as an *interconnection matrix*, and secondly, the matrix $\Omega(x,\dot{x})$ is the matrix of an *infinitesimal rotation.* In this Chapter, we focus on the second point of view, being interested in gyroscopic forces that prevent vehicles' collisions with obstacles or collisions with other vehicles.

The problem of avoiding obstacles is encountered in robotics and articulated vehicle systems. The objective is to build a control unit so that the target point is reached without collisions with obstacles during the movement. To achieve the desired movement, we use *potential forces, dissipative forces* and *gyroscopic forces.* The potential forces and the dissipative forces coordinate convergence to the target point, while gyroscopic forces coordinate it avoiding obstacles.

For simplicity, we judge in the plane x^1Ox^2. Suppose the vehicle is a unit mass point and the obstacle is the unit disk located at the origin. We want to design a feedback control law, capable of driving the vehicle (asymptotically) to the target point $x_T = (x_T^1, x_T^2)$, without attending to the obstacle. The *detection sensor* attached to the vehicle operates on a disk of radius r_{det} with the property that when the obstacle enters its range, the vehicle will have the reaction to change the direction of travel. The *safety sensor* has the role of preventing the collision.

The vehicle dynamics is described by a second order ODE of the form (Newton law)

$$\ddot{x}(t) = u(t),\ x = (x^1, x^2),\ u = (u^1, u^2),\ \ u = F_p + F_d + F_g + v,$$

where F_p is a conservative force which comes from a potential function $V : \mathbb{R}^2 \to \mathbb{R}$ having the minimum at target point x_T; F_d is a dissipative force; F_g is a gyroscopic force; v is an additional control force. For now on, we consider the force $v = 0$ and the other three forces of particular matrix form

$$F_p = -\nabla V(x),\ F_d = -D(x, \dot{x})\dot{x},\ F_g = \Omega(x, \dot{x})\dot{x},$$

where V is a potential function, the matrix D is symmetric and positive definite, and the matrix Ω is antisymmetric.

The explanations in Sections $1-4$ are based on paper [1]. Section 5 is based on the paper [10].

4.2 Avoidance of obstacles by gyroscopic forces

For simplicity, we accept

$$V(x) = \frac{1}{2}\,||x - x_T||^2,\ \ F_d = -2\dot{x},\ \ x = (x^1, x^2),\ \ \dot{x} = (\dot{x}^1, \dot{x}^2).$$

The total energy E of the vehicle is

$$E(x, \dot{x}) = \frac{1}{2}\,||\dot{x}||^2 + V(x). \tag{1}$$

Since

$$\frac{d}{dt}E(x, \dot{x}) = \langle \dot{x}, F_d \rangle = -2\,||\dot{x}||^2 \leq 0, \tag{2}$$

the total energy decreases in time (is dissipative).

Let's show that the vehicle does not collide with the obstacle for non-zero velocity, when

$$(3) \qquad E(x(0), \dot{x}(0)) \leq \frac{1}{2} V_{max}^2,$$

with V_{max} imposed.

Suppose the obstacle is convex. Then the vector $\vec{r}(x) = a(x)\vec{i} + b(x)\vec{j}$ determined by the vehicle position x and a point near the obstacle is well defined. Denote: $r(x) = \|\vec{r}(x)\|$ the *distance between the vehicle and the obstacle*; r_{det} *detection sensor radius*; $\det[\vec{r}(x), \dot{x}]$ the *orientation function.*

The mathematical model for movement (Newton Law, second order differential equation system) must reflect the following ideas: (1) the vehicle is deviated by gyroscopic force only when it detects an obstacle within the detection sensor range $r(x) \leq r_{det}$ and only if it is facing the obstacle, i.e., $\langle \vec{r}(x), \dot{x} \rangle > 0$; (2) the role of gyroscopic force is to rotate the velocity vector; (3) the sense of rotation depends on the orientation of vectors $\vec{r}(x)$ and $\dot{x}$, i.e., the sign of the determinant $\det[\vec{r}(x), \dot{x}]$. For that we introduce the function

$$\omega(x, \dot{x}) = \begin{cases} \frac{\pi V_{max}}{r(x)} & \text{if } r(x) \leq r_{det}, \langle \vec{r}(x), \dot{x} \rangle > 0, \det[\vec{r}(x), \dot{x}] \geq 0 \\ -\frac{\pi V_{max}}{r(x)} & \text{if } r(x) \leq r_{det}, \langle \vec{r}(x), \dot{x} \rangle > 0, \det[\vec{r}(x), \dot{x}] < 0 \\ 0 & \text{otherwise} \end{cases}$$

and using this function we define the gyroscopic force

$$F_g = \begin{bmatrix} 0 & -\omega(x, \dot{x}) \\ \omega(x, \dot{x}) & 0 \end{bmatrix} \dot{x}.$$

The sense of rotation is given by the sign of the function $\omega(x, \dot{x}) = \frac{\pi V_{max}}{r(x)}$. Then the vehicle movement is described by the second order differential system

$$(4) \qquad \ddot{x} = \begin{bmatrix} -2 & -\omega(x, \dot{x}) \\ \omega(x, \dot{x}) & -2 \end{bmatrix} \dot{x} - (x - x_T).$$

We rely on the following

Lemma 4.2.1. *If we denote*

$$\theta(t) = \int_{t_c - \Delta t}^{t} \omega(x(s), \dot{x}(s)) ds = \int_{t_c - \Delta t}^{t} \frac{\pi V_{max}}{r(x(s))} ds,$$

then the primitive of the matrix

$$\Omega(x,\dot{x}) = \begin{bmatrix} -2 & -\omega(x,\dot{x}) \\ \omega(x,\dot{x}) & -2 \end{bmatrix}$$

has the exponential

$$e^{-2t}\begin{bmatrix} \cos\theta(t) & -\sin\theta(t) \\ \sin\theta(t) & \cos\theta(t) \end{bmatrix} = e^{-2t}A(\theta(t)).$$

Proof. With the above notation, the primitive of the matrix $\Omega(x(t),\dot{x}(t))$ is

$$\begin{bmatrix} -2t & -\theta(t) \\ \theta(t) & -2t \end{bmatrix},$$

and its exponential has the form in Lemma. Indeed, if $B(t_1)$ commute with $B(t_2)$, for any t_1, t_2, then the differential equation of the form $\dot{x}(t) = B(t)x(t)$ has the solution $x(t) = \exp\left(\int B(t)dt\right)c$, where c is a constant vector. □

Theorem 4.2.2. *If the inequality (3) is satisfied, then the vehicle does not collide with the obstacle for nonzero velocity.*

Proof. Suppose the opposite, that is, the vehicle reaches the obstacle at the moment $t = t_c < \infty$ with speed $\dot{x}(t_c) \neq 0$. We take a small lap of time $\Delta t > 0$ and analyse the dynamics over time interval $I = [t_c - \Delta t, t_c^-]$. Without losing the generality, we can suppose $\det[\vec{r}(x), \dot{x}] \geq 0$ on I.

We transform the second order differential equation (4), in an integro-differential equation, in the sense of finding the velocity vector $\dot{x}(t)$, on the interval I:

$$\dot{x}(t_c^-) = e^{-2\Delta t}A(t_c^-)\dot{x}(t_c - \Delta t) - \int_{t_c-\Delta t}^{t_c^-} e^{-2(t_c^- - \tau)}A(\theta(t_c^-) - \theta(\tau))(x(\tau) - x_T)d\tau.$$

The norm $\|x(t)\|$ is bounded on I and the variation Δt is very small. Then the previous equality is approximated by

$$\dot{x}(t_c^-) = e^{-2\Delta t}A(t_c^-)\dot{x}(t_c - \Delta t) + \mathcal{O}(\Delta t).$$

Since the velocity vector field $\dot{x}(t)$ is continuous and $V(x) \geq 0$, we have $\beta < ||\dot{x}(t)|| \leq V_{max}$, $\forall t \in I$ and

$$V_{max}\,\Delta t \geq r(x(t_c - \Delta t)).$$

On the interval I, the trajectory approaches the obstacle. Therefore

$$r(x(t)) \leq r(x(t_c - \Delta t)),\ t \in I.$$

It follows $\theta(t_c^-) \geq \pi$, independent of variation Δt.

In conclusion, the velocity vector field $\dot{x}(t)$ rotates with an angle, let's say, bigger than $\frac{3\pi}{4}$, on the interval $I = [t_c - \Delta t, t_c^-]$, for Δt small. On the other hand, due to the assumption of collision at the moment $t = t_c$, with nonzero velocity, the velocity cannot rotate so much because of its continuity. For example, we can take Δt such that $\theta(t_c^-) \leq \frac{\pi}{2}$. We came to a contradiction and as a result there is no collision between vehicle and obstacle, in finite time, at nonzero velocity. □

4.3 Two types of collisions

The previous Section showed that there is no finite time collision, at non-zero velocity. On the other hand, a collision can take place either in finite time, but at zero velocity, or in infinite time. The finite time case can only occur if $\dot{x}(t_c) = 0$, where t_c is the moment of collision. The one in infinite time implies the existence of an increasing sequence $t_i \to \infty$ such that the sequence of points $x(t_i)$ converges to the obstacle. From the relations (1) and (2), we deduce

$$\int_0^\infty ||\dot{x}(\tau)||^2\, d\tau \leq \frac{1}{2}\, E(0) < \infty.$$

In conclusion, there exists a sequence s_i that increases to ∞ such that

$$\lim_{i \to \infty} \dot{x}(s_i) = 0.$$

This means that the vehicle slows down at least sporadically.

The two types of collisions are called zero velocity collisions. We can enter a control v to reach a target or to avoid zero velocity collisions. For example, a control that maintains a minimum speed eliminates collisions. This idea will be presented in Section 5.

4.4 Asymptotic convergence of the vehicle to a target

We move the explanations in the phase space $\mathbb{R}^4$. Suppose that the trajectory $(x(t), \dot{x}(t))$ satisfies the relation (3) and it doesn't end as a zero velocity collision. Since the point $x(t)$ is at a certain distance from the obstacle, there is an open set $W \subset \mathbb{R}^2$ containing the obstacle and so that the trajectory is in the compact set

$$K = E^{-1}([0, E(t = 0)]) \setminus (W \times \mathbb{R}^2).$$

Then the trajectory $x(t)$ exists for all $t \geq 0$. The asymptotic convergence to the target $(x_T, 0)$ can be obtained using the energy as a Lyapunov function.

We now prefer another Lyapunov function which gives an alternative proof of convergence. For $0 < \varepsilon < 1$, we consider the function

$$U(x, \dot{x}) = E(x, \dot{x}) + \varepsilon dV(x)(\dot{x})$$

$$= \frac{1}{2}\left((\dot{x}^1)^2 + (\dot{x}^2)^2 + (x^1 - x_T^1)^2 + (x^2 - x_T^2)^2\right)$$

$$+\varepsilon\left((x^1 - x_T^1)\dot{x}^1 + (x^2 - x_T^2)\dot{x}^2\right)$$

which plays the role of Lyapunov function. Indeed, the function U checks the conditions: (1) $U(x_T, 0) = 0$ and $U(x, \dot{x}) > 0$ on $K \setminus \{(x_T, 0)\}$; (2) the point $(x_T, 0)$ is the only critical point of the function U on K; (3) along an arbitrary trajectory we have

$$\frac{dU}{dt} = -(2 - \varepsilon)||\dot{x}||^2 - \varepsilon||x - x_T||^2 - 2\varepsilon\langle x - x_T, \dot{x}\rangle$$

$$+\varepsilon(x - x_T)\begin{bmatrix} 0 & -\omega(x, \dot{x}) \\ \omega(x, \dot{x}) & 0 \end{bmatrix}\dot{x}.$$

Since the function $\omega(x, \dot{x})$ is bounded on the set K, we can find $\varepsilon > 0$ and $c > 0$ such that

$$\frac{dU}{dt} \leq -c\, U \leq 0,$$

on the set K. It follows that $U(t) \leq U(0)e^{-ct}$. This relation shows that any trajectory converges asymptotically to a target.

4.5 Optimal control problem

In this Section, the analysis is focused on single-time optimal control problems based on simple integral cost functional from a first order Lagrangian and the second order ODEs system that describe a controlled movement of the vehicle. The basic topics of our theory include: variational differential systems, adjoint differential systems, single-time maximum principle.

Vehicle dynamics control are able to improve the handling and stability performance of vehicles to a good extent.

Taking $u(t) = (u^1(t), u^2(t))$ like a control, we can imagine the following optimal control problem (a least squares problem): *find*

$$\min_{(u^1,u^2)} \frac{1}{2}\int_0^T (\|\dot{x}(t)\|^2 + \|u(t)\|^2)\, dt$$

subject to the controlled evolution ODE

$$\ddot{x} = \begin{bmatrix} -2 & -\omega(x,\dot{x}) \\ \omega(x,\dot{x}) & -2 \end{bmatrix} \dot{x} - (x - x_T) + u(t),$$

where $\omega(x(t),\dot{x}(t)) = \frac{\pi V_{max}}{r(x(t))}$, $u(t) \in \mathcal{U}, \forall t \in [0,T]$, *with the conditions*

$$x(0) = x_0,\ x(T) = x_T,\ \dot{x}(0) = \tilde{x}_0,\ \dot{x}(T) = \tilde{x}_T.$$

If we want the vehicle to only pass through the target point, then we ask that $\tilde{x}_T \neq 0$. In this case, we can impose $\|\dot{x}\| \geq a > 0$.

If we want the vehicle to stop at the target point, then $\tilde{x}_T = 0$. In this case, we ask the control to ensure that $\|\dot{x}\| \geq a > 0$, for any point x outside a neighbourhood of $\tilde{x}_T$.

We are looking for necessary conditions of optimality (for a pair (x,u)) in the previous optimal control problem. Setting

$$X(x(t),z(t),u(t)) = \begin{bmatrix} -2 & -\omega(x(t),z(t)) \\ \omega(x(t),z(t)) & -2 \end{bmatrix} z(t)$$

$$-(x(t) - x_T) + u(t),$$

we remark that the previous differential system can be rewritten as follows

$$\dot{x}(t) = z(t),\ \ \dot{z}(t) = X(x(t),z(t),u(t)).$$

We build the control Hamiltonian

$$H(x(t),z(t),u(t),p(t),q(t)) = \frac{1}{2}(\|z(t)\|^2 + \|u(t)\|^2)$$

$$+\langle p(t),z(t)\rangle + \langle q(t),X(x(t),z(t),u(t))\rangle.$$

Considering the differentiable variations $x(t,\epsilon), z(t,\epsilon)$, fulfilling

$$\dot{x}(t,\epsilon) = z(t,\epsilon), \dot{z}(t,\epsilon) = X(x(t,\epsilon), z(t,\epsilon), u(t)),$$

$x(t,0) = x(t), z(t,0) = z(t)$, and setting $x_\epsilon(t,0) = y(t), z_\epsilon(t,0) = v(t)$, we obtain the *variational differential system* (the summation over the repeated indices is assumed)

$$\dot{y}(t) = v(t),\ \ \dot{v}(t) = X_{x^j}(x(t), z(t), u(t))y^j(t) + X_{z^j}(x(t), z(t), u(t))v^j(t)$$

and the *adjoint differential system*

$$\dot{p}_j(t) = -\langle q(t), X_{x^j}(x(t), \dot{x}(t), u(t))\rangle,$$

$$\dot{q}_j(t) = -p_j(t) - \langle q(t), X_{\dot{x}^j}(x(t), \dot{x}(t), u(t))\rangle.$$

Theorem 4.5.1 (Simplified single-time maximum principle). *Let $(x, \hat{u})$ be an optimal pair. Then there exist a C^1 co-state variable $p = (p_i)$, respectively a C^2 co-state variable $q = (q_i)$, defined over $[0, T]$, such that:*

$$\dot{x}^j(t) = \frac{\partial H}{\partial p_j}(x(t), \dot{x}(t), \hat{u}(t), p(t), q(t)),\ \ \ddot{x}^j = \frac{\partial H}{\partial q_j}(x(t), \dot{x}(t), \hat{u}(t), p(t), q(t))$$

$$\forall t \in [0, T],\ j = 1, 2,\ x(0) = x_0,\ \dot{x}(0) = \tilde{x}_0;$$

the functions $p = (p_i), q = (q_i)$ satisfy the terminal problems

$$\dot{p}_j(t) = -H_{x^j}(x(t), \dot{x}(t), \hat{u}(t), p(t), q(t)),\ p_j(T) = 0,$$

$$\dot{q}_j(t) = -H_{\dot{x}^j}(x(t), \dot{x}(t), \hat{u}(t), p(t), q(t)),\ q_j(T) = 0;$$

the critical point conditions are

$$H_{u^i}(x(t), \dot{x}(t), \hat{u}(t), p(t), q(t)) = 0,\ \forall t \in [0, T],\ i = 1, 2,$$

and

$$\frac{\partial H}{\partial x^j}(x(t), \dot{x}(t), \hat{u}(t), p(t), q(t))$$

$$-\frac{d}{dt}\left[\frac{\partial H}{\partial \dot{x}^j}(x(t), \dot{x}(t), \hat{u}(t), p(t), q(t)) - p_j(t)\right] + \frac{d^2}{dt^2}[-q_j(t)] = 0,$$

$\forall t \in [0, T]$.

In our case,

$$H_{u^i} = u^i(t) + q_i(t) = 0$$

and the critical point $u(t) = -q(t)$ is a minimum point of the Hamiltonian H.

4.6 Self-evaluation problems

Problem 40. *The equations of motion (of a bicycle, motorcycle) can be written in matrix form*

$$M\ddot{q} + C\dot{q} + Kq = f,$$

where M is the symmetric matrix of the masses (terms containing the mass or geometry of the vehicle); C is the damping matrix, antisymmetric, even if an idealized vehicle has no dissipation (terms that contain the velocity of advancement v); K is the stiffness matrix (terms including gravitational constant g and square of velocity v^2); q is the vector of tilt and turn angles; f is external forces vector (moments). Comment on the existence of the solutions of the form $q = ue^{\lambda t}$ for the homogeneous equation and comment on the stability of the equilibrium positions.

Problem 41. *The vibrations of a pendulum with viscous linear damping are described by the differential equation*

$$m\ell\ddot{\varphi} + \gamma\ell\dot{\varphi} + mg\sin\varphi = 0,$$

where φ is the angle of the pendulum relative to the vertical, m is the mass of the pendulum, ℓ is the length of the pendulum, g is the gravitational acceleration, and γ is the damping index. Rewrite as a first order differential system, find the equilibrium positions and comment on their stability.

Problem 42. *Let $x = (x^j) \in \mathbb{R}^n$. A vector field $X(x)$ on $\mathbb{R}^n$, of components $X^i(x), i = 1, ..., n$, generates a gyroscopic force $F(x, \dot{x})$ of components*

$$F^i = \left(\frac{\partial X^i}{\partial x^j} - \frac{\partial X^j}{\partial x^i}\right)\dot{x}^j,$$

with summation over the index j. Find the gyroscopic force F in the case of Euler vector field $(ax^2x^3, bx^3x^1, cx^1x^2)$.

Problem 43 (Continuation). *Find the gyroscopic force F induced by the Lotka-Volterra vector field*

$$X^i = \lambda^i x^i(1 - a^i_j x^j).$$

Problem 44 (Continuation). *Determine the gyroscopic force F induced by the dynamical system*

$$\dot{x}_i = 2\left(c_i x_i - \sum_{j=1}^{n} c_j x_j^2\right) x_i,\; x_i \geq 0,\; \sum_{i=1}^{n} x_i = 1,\; i, j = 1, ..., n.$$

Bibliography

[1] D. E. Chang, J. E. Marsden, *Gyroscopic forces and collision avoidance with convex obstacles*, W. Kang et al. (Eds.): New Trends in Nonlinear Dynamics and Control, LNCIS 295, pp. 145-159, Springer-Verlag Berlin Heidelberg, 2003.

[2] C.-K. Chen, T.-D. Chu, X.-D. Zhang, *Modeling and control of an active stabilizing assistant system for a bicycle*, Sensors (mdpi) 19, 248 (2019), pp. 1-21.

[3] C. Corcodel, C. Udriste, *Evolution metrics and geometric vector fields*, ICOSSSE-10 Proceedings of the 9th WSEAS International Conference on System Science and Simulation in Engineering, 2010, 59-64.

[4] D. Glickenstein, *Introduction to Flows on Riemannian Metrics*, Math. 538, Springer, 2009.

[5] R. S. Hamilton, *Three-manifolds with positive Ricci curvature*, J. Differ. Geom., **17**(1982), No. 2, 255-306.

[6] D.- X. Kong, K. Liu, *Wave character of metrics and hyperbolic geometry*, http://www.cms.zju.edu.cn/UploadFiles /AttachFiles/200682885946597.pdf

[7] B. Mashadi, M. Gowdini, *Vehicle dynamics control by using an active gyroscopic device*, Journal of Dynamic Systems, Measurement, and Control, 137 (2015), 121007-1 to 121007-12.

[8] J. Morgan and G. Tian, *Ricci Flow and the Poincaré Conjecture*, Clay Math. Monographs, Vol. 3, Amer. Math. Soc. and Clay Math. Institute, 2007.

[9] N. Sheridan, *Hamilton's Ricci Flow*, Honours Thesis, The University of Melbourne, Department of Mathematics and Statistics, November 2006.

[10] S. Treanta, C. Udriste, *Optimal control problems with higher order ODEs constraints*, Balkan Journal of Geometry and Its Applications, 18, 1 (2013), pp. 71-86.

[11] C. Udriste, M. Toda, *Volume dependence on the Riemannian metric*, Sci. Bull. PUB, Series A: Appl. Math. Phys. **55**(1993), No. 3-4, 285-298.

[12] C. Udriste, M. Toda, *Influence of the evolution Riemannian metrics on the volume*, in "Proc. 23rd Conf. Geom. Topol., D. Andrica et al. (Eds.), Cluj-Napoca, 1994, pp. 194-202.

[13] C. Udriste, *Field Lines*, Technical Publishing House, Bucharest, 1988 (in Romanian).

[14] C. Udriste, *Geometric Dynamics*, Mathematics and Its Applications, 513, Kluwer Academic Publishers, Dordrecht, Boston, London, 2000.

[15] C. Udriste, L. Matei, *Lagrange-Hamilton Theories* (in Romanian), Monographs and Textbooks 8, Geometry Balkan Press, Bucharest, 2008.

Chapter 5

NAMBU DYNAMICAL SYSTEMS

Motto:

"I can't recall them all and tell them all
But I would tell everything I know, but to whom?
Because, as a child I fought
In turn under King Tempest
And I also knew the Rabid King
Just as his land."

George Coşbuc - *Fulger's Death*

Nambu theory refers to special dynamical systems, defined by functional determinants. Among other things, we make a connection between Nambu dynamical systems and gradient dynamical systems. Part of the theory is presented in a three-dimensional space, but some results can be transposed on any dimension (even or odd). Suggestions for further reading: [1]-[12].

5.1 Nambu dynamical systems

In Hamiltonian mechanics a smooth manifold M is called a *Poisson manifold*, and the *functions ring* $A = C^\infty(M)$ is called *algebra of observables*, if there exists a map $\{\ ,\ \} : A \otimes A \to A$ (Poisson bracket) with the following properties:

(i) skew-symmetry

$$\{f_1, f_2\} = -\{f_2, f_1\},$$

(ii) Leibniz rule (derivation property)

$$\{f_1 f_2, f_3\} = f_1\{f_2, f_3\} + f_2\{f_1, f_3\},$$

(iii) Jacobi identity

$$\{f_1, \{f_2, f_3\}\} + \{f_3, \{f_1, f_2\}\} + \{f_2, \{f_3, f_1\}\} = 0,$$

for all $f_1, f_2, f_3 \in A$.

We start with the simplest phase space for Hamiltonian mechanics - a two-dimensional Euclidean space $\mathbb{R}^2$, with coordinates x, y and the canonical Poisson bracket $\{f_1, f_2\} = \frac{\partial(f_1,f_2)}{\partial(x,y)}$. This bracket leads to a Hamilton's equation of motion $\frac{d}{dt}f(x(t), y(t)) = \{H, f\}(x(t), y(t))$, where f is a classical observable - a smooth function on the phase space and H is a Hamiltonian.

The canonical Nambu bracket is defined for a triple of classical observables on the 3-dimensional phase space $\mathbb{R}^3$ with coordinates x, y, z by the following beautiful formula

$$\{f_1, f_2, f_3\} = \frac{\partial(f_1, f_2, f_3)}{\partial(x, y, z)},$$

where the right-hand side stands for the Jacobian of the mapping

$$f = (f_1, f_2, f_3) : \mathbb{R}^3 \to \mathbb{R}^3.$$

This formula naturally generalizes the usual Poisson bracket from binary to ternary operation on classical observables. Generalized Nambu-Hamilton equations of motion involve two "Hamiltonians" H_1 and H_2 and have the form

$$\frac{d}{dt}f(x(t), y(t), z(t)) = \{H_1, H_2, f\}(x(t), y(t), z(t)).$$

The corresponding phase flow in the phase space is divergence-free and preserves the standard volume form $dx \wedge dy \wedge dz$ - analog of the Liouville theorem for Nambu mechanics.

The generalization of these examples leads to a concept of *Poisson manifolds* - smooth manifolds endowed with a Poisson bracket structure satisfying skew-symmetry condition, Leibniz rule and Jacobi identity.

Now let us explain the main ideas of this theory in more details. The dynamical systems proposed by Nambu generalize Hamilton's ideas on the space $\mathbb{R}^n$, where the dimension n is odd or even. The vector variable is $x = (x^1, ..., x^n) \in \mathbb{R}^n$. In Nambu's conception, the evolution of the dynamic variable $F(x)$ is fixed by $n-1$ Hamiltonian functions $H_1, ..., H_{n-1}$, by the Jacobi determinant (the determinant of the Jacobian matrix)

$$\frac{\partial(F, H_1, ..., H_{n-1})}{\partial(x^1, ..., x^n)}$$

and by the differential equation

$$\frac{d}{dt}F(x(t)) = \frac{\partial(F, H_1, ..., H_{n-1})}{\partial(x^1, ..., x^n)}(x(t)). \tag{1}$$

Generally, the C^∞ functions $F_1(x), ..., F_n(x)$ define the *Nambu bracket*

$$\{F_1, ..., F_n\} = \frac{\partial(F_1, F_2, ..., F_n)}{\partial(x^1, ..., x^n)}.$$

This bracket is anti-symmetric and works like a derivation on the space of functions. If we use the coordinate functions x^i instead of the dynamic variable $F(x)$, then it appears the differential system

$$\frac{dx^i}{dt}(t) = \frac{\partial(x^i, H_1, ..., H_{n-1})}{\partial(x^1, ..., x^n)}(x(t)),\ i = 1, ..., n \tag{2}$$

of type Nambu. Since the Hamiltonians H_k, $k = 1, ..., n-1$, are first integrals of this system, i.e., $\frac{d}{dt}H_k = 0$, the system (2) has the implicit solution $H_k(x) = c_k$, $k = 1, ..., n-1$. Also from the differential system (2), we find the Liouville condition (with a sum over the index $i = 1, ..., n$)

$$\frac{\partial}{\partial x^i}\frac{dx^i}{dt} = 0. \tag{3}$$

Though this paragraph poses more questions, we feel that the subject of Nambu theory might be relevant for future development of mathematical and engineering students.

5.2 Gradient dynamical system

Let $\Phi : \mathbb{R}^n \to \mathbb{R}$ be a C^∞ scalar field. A dynamical system of the form

$$\frac{dx^i}{dt}(t) = -\frac{\partial \Phi}{\partial x^i}(x(t)) \tag{4}$$

on $\mathbb{R}^n$, is called (minus) *gradient system*. Here, the vector $\frac{dx^i}{dt}(t)$ tangent to the curve solution $x^i = x^i(t)$ is orthogonal to the constant level hypersurface $\Phi(x) = c$ (if $\nabla\Phi \neq 0$). Also, the time evolution $\frac{d\Phi}{dt} = -||\nabla\Phi||^2$ of the function Φ shows that it decreases along the trajectory, reaching the minimum for $t \to \infty$, if the limit exists.

Remark 5.2.1. *If $n = 2$ (the space is two-dimensional), then the differential system (4) is Hamiltonian if the functions of Hamilton K and Φ are the real and imaginary part of a complex monogenic (analytic) function. In fact, these functions must satisfy the Cauchy-Riemann equations of monogeneity*

$$\partial_1 K = \partial_2 \Phi,\ \partial_2 K = -\partial_1 \Phi.$$

One of our aim is to detect the conditions in which a dynamical system of Nambu type is a gradient system. This can only be done in particular cases (for example, on $\mathbb{R}^3$).

5.3 Nambu-gradient systems in space

Let's consider $x = (x^1, x^2, x^3) \in \mathbb{R}^3$. In this case, the Nambu system can be written either in the form

$$\dot{x}^i(t) = \frac{\partial(H, G)}{\partial(x^j, x^k)}(x(t)),$$

where (i, j, k) is a cyclic permutation of $(1, 2, 3)$, or in the Euler form

$$\frac{dx}{dt}(t) = \nabla H(x(t)) \times \nabla G(x(t)).$$

Since the functions (Euler potentials) H are G first integrals, the orbits of such a differential system have the implicit equations $H(x) = c_1,\ G(x) = c_2$.

To write a determinant using its elements, we need the permutation symbol (signature)

$$\epsilon^{ijk} = \begin{cases} 1 & if \quad (i,j,k) \in \{(1,2,3),(2,3,1),(3,1,2)\} \\ -1 & if \quad (i,j,k) \in \{(1,3,2),(3,2,1),(2,1,3)\}. \end{cases}$$

Then the Nambu bracket (functional determinant) can be written

$$\{f, g, h\} = \varepsilon^{ijk}\, \partial_i f\, \partial_j g\, \partial_k h,\ \forall f, g, h \in C^\infty(\mathbb{R}^3).$$

Consequently, a Nambu differential system fixed by the functions H, G is transcribed as $\dot{x}^i = \{x^i, H, G\}$.

Let's investigate if this Nambu system is reduced to a gradient system

$$\frac{dx}{dt}(t) = -\nabla\Phi(x(t)),$$

i.e., there exists a function Φ such that

$$\nabla H \times \nabla G = -\nabla\Phi.$$

This relation shows that the three functions H, G, Φ determine a local coordinate system on $\mathbb{R}^3$, except for the equilibrium points of the dynamical system (where the constant-level surfaces of the functions H, G are tangent).

The Liouville condition (3) implies $\Delta\Phi = \nabla^2\Phi = 0$, i.e., the function Φ is harmonic. The following questions arise: (1) given the functions H, G, which conditions determine the function Φ? (2) given the function Φ, which conditions determine the functions H, G?

Let us take $H = a_i x^i$ (linear form), $G = \frac{1}{2} b_{ij} x^i x^j$ (quadratic form). Then the function Φ exists only if

$$a_1 b_{12} - a_2 b_{11} = 0$$

$$a_1 b_{13} - a_2 b_{23} - a_3(b_{11} + b_{22}) = 0$$

$$a_1(b_{22} + b_{33}) - a_2 b_{12} - a_3 b_{13} = 0.$$

In this case, the function Φ is the solution of the completely integrable PDE system

$$\frac{\partial\Phi}{\partial x^1} = (a_3 b_{2j} - a_2 b_{3j})x^j, \ \frac{\partial\Phi}{\partial x^2} = (a_1 b_{3j} - a_3 b_{1j})x^j, \ \frac{\partial\Phi}{\partial x^3} = (a_2 b_{1j} - a_1 b_{2j})x^j.$$

Obviously, not every Nambu system is a gradient system. A Nambu system $\frac{dx^i}{dt} = \frac{\partial(H,G)}{\partial(x^j,x^k)}$, where (i, j, k) is a cyclic permutation, is a gradient system if and only if the function Φ in the relation $\frac{\partial(H,G)}{\partial(x^j,x^k)} = -\frac{\partial\Phi}{\partial x^i}$ satisfies the conditions of complete integrability $\frac{\partial^2\Phi}{\partial x^l \partial x^i} = \frac{\partial^2\Phi}{\partial x^i \partial x^l}$, that is, the relationships

$$\frac{\partial}{\partial x^l}\frac{\partial(H,G)}{\partial(x^j,x^k)} = \frac{\partial}{\partial x^i}\frac{\partial(H,G)}{\partial(x^j,x^k)}$$

are identically satisfied. These conditions can be transcribed

$$\nabla(\partial_i G \ \nabla H - \partial_i H \ \nabla G) = 0.$$

It follows that for each $i = 1, 2, 3$, there exists one vector field A_i such that

$$\partial_i G \ \nabla H - \partial_i H \ \nabla G = \nabla \times A_i, \ i = 1, 2, 3.$$

Adding these three relationships, introducing the vector $U = (1, 1, 1)$ and denoting $D = A_1 + A_2 + A_3$, we deduce

$$\langle U, \nabla G\rangle \nabla H - \langle U, \nabla H\rangle \nabla G = \nabla \times D \tag{5}$$

and by $\nabla H \times \nabla G = -\nabla\Phi$ we find

$$D = \Phi U + \nabla F, \tag{6}$$

where F is an arbitrary function. Thus, the conditions on the function H, G are fixed by the relations (5), and the vector field D is given by (6).

Conversely, being given the field Φ, the relation $\nabla H \times \nabla G = -\nabla\Phi$ defines

$$\nabla H = \frac{\nabla G \times \nabla \Phi}{||\nabla G||^2} + \Lambda \nabla G,$$

where Λ is an arbitrary function. However, we must ensure that the functions G and Λ are chosen so that the conditions of complete integrability to be satisfied.

The conditions of complete integrability, after eliminating two of the partial derivatives of Λ, lead to a nonlinear second order equation for G, completely determined by giving Φ and the remaining partial derivative of Λ. If $G = g(x^3)$, then it rests the condition

$$\left(\frac{\partial^2 \Phi}{\partial {x^1}^2} + \frac{\partial^2 \Phi}{\partial {x^2}^2}\right)\left(\frac{dg}{dx^3}\right) = 0.$$

5.4 Stability of equilibrium points

The equilibrium points of a gradient-type Nambu dynamical system are the solutions of the algebraic system $\nabla H(x) \times \nabla G(x) = -\nabla\Phi(x) = 0$. Let x_0 be such a point. A Lyapunov function V implies the stability if in a neighbourhood of this point we have $V(x_0) = 0$, $V(x) > 0$, $x \neq$

x_0, $\frac{d}{dt}V(x(t)) \leq 0$. If $\frac{d}{dt}V(x(t)) < 0$, then the stability is asymptotic. The relation

$$\frac{d}{dt}V(x(t)) = -\langle \nabla V, \nabla \Phi \rangle$$

shows that if x_0 is an isolated minimum point of the function Φ, then $V(x) = \Phi(x) - \Phi(x_0)$ is a Lyapunov function and the point x_0 is asymptotically stable.

For a gradient system $\frac{dx}{dt}(t) = -\nabla\Phi(x(t))$, the matrix of linearization $M = (M_{ij}) = (-\partial_i\partial_j\Phi(x_0))$, around the point x_0, obtained from $-\nabla\Phi$, is symmetric and hence the eigenvalues are real. Conditions on the G and H functions appear. Liouville's condition implies $tr(M_{ij}) = 0$, because Φ is harmonic. The Jacobian matrix associated with $\nabla H \times \nabla G$ is $L = M$.

Example Let's consider $\Phi = -x^1x^2x^3$, $2H = {x^2}^2 - {x^1}^2$, $2G = {x^3}^2 - {x^1}^2$. By calculation it is verified $\nabla H \times \nabla G = -\nabla\Phi$. The equilibrium points are given as $x^1 = x^2 = 0$ etc.

5.5 The $SU(n)$ - isotropic harmonic oscillator

The Hamiltonian for the n-dimensional simple harmonic oscillator is a very important example in quantum mechanics. If we set to one all the frequencies, then this Hamiltonian can be written $H = \sum_{i=1}^{n}(p_i^2 + q_i^2)$. It is invariant under the symmetry group $SU(n)$ (the special unitary group of degree $n \equiv$ the Lie group of $n \times n$ unitary matrices with determinant 1). The associated dynamical system $\dot{p}_i = -\frac{\partial H}{\partial q_i} = -2q_i$, $\dot{q}_i = \frac{\partial H}{\partial p_i} = 2p_i$, $i = 1, ..., n$, has two kind of first integrals: (1) the angular momenta $L_{ij} = q_ip_j - q_jp_i$, $i, j = 1, ..., n$, in number of $\frac{n^2-n}{2}$; (2) $A_{ij} = p_ip_j + q_iq_j$, $i, j = 1, ..., n$, in number of $\frac{n^2+n}{2}$. The diagonal components A_{ii} are the individual energies associated with the separate one-dimensional oscillations. The off-diagonal components $A_{ij}, i \neq j$, provide the hidden first integrals of motion.

In other words, the L_{ij}'s and the A_{ij}'s, in number of n^2, provide $2n-1$ independent first integrals of motion within the $2n$-dimensional phase space, and therefore all orbits lie on one-dimensional manifolds (curves). Explicitly, these orbits are closed curves (circles).

Two-dimensional case One defines the following functions

$$S_1 = \frac{1}{2}(A_{12} + A_{21}),\ S_2 = \frac{1}{2}(A_{22} - A_{11}),\ S_3 = \frac{1}{2}L_{12}.$$

We can verify that $\{S_i, S_j\} = \epsilon_{ijk} S_k$, where ϵ_{ijk} is a cyclic permutation of $\{1, 2, 3\}$, which are simply the commutation relations for the Lie algebra $su(2)$. Furthermore,

$$S_1^2 + S_2^2 + S_3^2 = S^2 = \frac{1}{4} H^2.$$

The constant level sets of the first integrals of motion for this system are written in the form

$$I_1 : p_1^2 + q_1^2 = c_1, I_2 : p_2^2 + q_2^2 = c_2, I_3 : q_1 p_2 - q_2 p_1 = c_3, I_4 : p_1 p_2 + q_1 q_2 = c_4.$$

Using I_1, I_2, I_3 as the independent generalized Hamiltonians, we can describe a corresponding Nambu dynamics as follows. Define the following Nambu bracket,

$$\{f, I_1, I_2, I_3\} = \frac{-1}{2c_4} \frac{\partial(f, I_1, I_2, I_3)}{\partial(p_1, p_2, q_1, q_2)},$$

where f is an arbitrary function on the phase space coordinates. Then, the associated dynamical system is

$$\dot{p}_1 = \{p_1, I_1, I_2, I_3\} = -2q_1, \text{ etc.}$$

Another definition for Nambu bracket, using the independent Hamiltonians I_1, I_2, I_4, and giving also the correct equations of motion is

$$\{f, I_1, I_2, I_4\} = \frac{1}{2c_3} \frac{\partial(f, I_1, I_2, I_4)}{\partial(p_1, p_2, q_1, q_2)}.$$

There are many other ways to build Nambu dynamics. This extra flexibility not found in Hamiltonian mechanics is obviously due to the multiple Hamiltonian structure of Nambu mechanics. Finally, it is relatively straightforward to extend the above arguments to higher dimensional systems. One simply uses a part of the first integrals given by L_{ij} and A_{ij} as the independent generalized Hamiltonians and finds the correct constants related to these integrals to multiply the Jacobian term in the definition of the Nambu bracket. Thus, the $SU(n)$ - isotropic harmonic oscillator is realizable as an n-th order Nambu mechanical system.

5.6 The $SO(4)$-Kepler Problem

$SO(4)$ is a special orthogonal group (rotation group). It consists of all orthogonal 4×4 matrices of determinant 1.

On $\mathbb{R}^3 = \{(x,y,z)\}$, let $r = \sqrt{x^2+y^2+z^2}$. The well known Kepler Hamiltonian is $H = \frac{1}{2}\|p\|^2 - \frac{1}{r}$. Because H possesses rotational symmetry, the orbital angular momentum $\vec{L} = \vec{r} \times \vec{p}$ is a first integral of motion. This rotational symmetry implies that the orbit lies in some two dimensional plane, though it is not enough to ensure that the orbit is closed. An extra dynamical symmetry must exist for a closed orbit since the first integrals of motion H and $\vec{L}$ only reduce the phase space to a two dimensional (as opposed to a one dimensional) manifold. Such a first integral was discovered by Laplace (but it is called the Runge-Lenz vector in classical mechanics or the Lenz-Pauli vector in quantum mechanics) and is given by $\vec{A} = \vec{p} \times \vec{L} - \frac{1}{r}\vec{r}$. One checks that $\{A_i, L_j\} = \epsilon_{ijk}A_k$ and

$$\{A_i, A_j\} = -\epsilon_{ijk}\left(\|p\|^2 - \frac{2}{r}\right)L_k = -2HL_k = -2EL_k,$$

where E is the constant value taken by H and ϵ_{ijk} is a cyclic permutation of $\{1,2,3\}$. For bound state problems ($E < 0$), one can define a new conserved vector $\vec{D} = \frac{1}{\sqrt{-2E}}\vec{A}$. Then, one finds that the commutation relations reduce to

$$\{L_i, L_j\} = \epsilon_{ijk}L_k, \ \{D_i, L_j\} = \epsilon_{ijk}D_k, \ \{D_i, D_j\} = \epsilon_{ijk}L_k.$$

which is the Lie algebra $so(4)$ (note that for scattering problems where $E > 0$, one instead finds the Lie algebra $so(3,1)$). Explicitly, we have $H = \frac{1}{2}\|p\|^2 - \frac{1}{\|q\|}$, which shows that the Kepler motion is generated by a flow and the geometry of the manifold as primordial data (see [11]). Other first integrals are

$$I_1 : p_2(q_1p_2 - q_2p_1) - p_3(q_3p_1 - q_1p_3) - \frac{q_1}{\|q\|} = c_1,$$

$$I_2 : p_3(q_2p_3 - q_3p_2) - p_1(q_1p_2 - q_2p_1) - \frac{q_2}{\|q\|} = c_2,$$

$$I_3 : p_1(q_3p_1 - q_1p_3) - p_2(q_2p_3 - q_3p_2) - \frac{q_3}{\|q\|} = c_3,$$

$$I_4 : q_2p_3 - q_3p_2 = c_4, \ I_5 : q_3p_1 - q_1p_3 = c_5, \ I_6 : q_1p_2 - q_2p_1 = c_6.$$

All the seven first integrals satisfy two relations

$$I_1^2 + I_2^2 + I_3^2 = 1 + 2H(I_4^2 + I_5^2 + I_6^2), \ I_1I_4 + I_2I_5 + I_3I_6 = 0.$$

Therefore, there exists only five independent constants of motion as expected. As it was the case for the harmonic oscillator, one may choose any five of the above six I_i's (or products thereof) as the generalized Hamiltonians. For example, one may define the Nambu bracket for this system as

$$\{f, I_2, I_3, I_4, I_5, I_6\} = \frac{1}{c_4(c_4^2 + c_5^2 + c_6^2)} \frac{\partial(f, I_2, I_3, I_4, I_5, I_6)}{\partial(p_1, p_2, p_3, q_1, q_2, q_3)}$$

or

$$\{f, I_1, I_3, I_4, I_5, I_6\} = \frac{-1}{c_5(c_4^2 + c_5^2 + c_6^2)} \frac{\partial(f, I_1, I_3, I_4, I_5, I_6)}{\partial(p_1, p_2, p_3, q_1, q_2, q_3)},$$

etc.

Open problem Try to incorporate the original Hamiltonian H into a form of the Nambu bracket.

Example 5.6.1. Let us consider the Euler or Lagrange system (sometimes called Nahm's system in the theory of static $SU(2)$ - monopoles), which is given on $\mathbb{R}^3 = \{(x^1, x^2, x^3)\}$ by the following equations of motion

$$\dot{x}^1 = x^2x^3,\ \dot{x}^2 = x^3x^1,\ \dot{x}^3 = x^1x^2.$$

The differential equations can be written in Nambu form

$$\dot{x}^i = \{H_1, H_2, x^i\}, i = 1, 2, 3,$$

where $H_1 = {x^1}^2 - {x^2}^2$, $H_2 = {x^1}^2 - {x^3}^2$. The constant level surfaces (hyperbolic cylinders) of H_1 and H_2 confine the phase flow to the intersection of two quadrics in $\mathbb{R}^3$ - a locus of an elliptic curve (so that the system can be integrated by elliptic functions).

The previous dynamical system is potential. The global potential is $f : \mathbb{R}^3 \to \mathbb{R}, f(x^1, x^2, x^3) = x^1x^2x^3$. The surfaces of nonzero constant level attached to f are Tzitzeica surfaces. The axes Ox^1, Ox^2, Ox^3 consists of equilibrium points.

From the symmetric differential system

$$\frac{dx^1}{x^2x^3} = \frac{dx^2}{x^3x^1} = \frac{dx^3}{x^1x^2}$$

we find the Cartesian implicit equations of the family of field lines

$${x^1}^2 - {x^2}^2 = c_1,\ {x^1}^2 - {x^3}^2 = c_2.$$

Suppose that the motion of a particle in $\mathbb{R}^3$ has the trajectories of the previous dynamical system.

(i) If two of the numbers $x^1(0), x^2(0), x^3(0)$ are zero, then the particle does not move.

(ii) If $x^1(0) = x^2(0) = 1, x^3(0) = 0$, then the trajectory of the motion has the parametric equations

$$x^1(t) = \sec t,\ x^2(t) = \sec t,\ x^3(t) = \tan t,\ t \neq (2k+1)\frac{\pi}{2},\ k \in \mathbb{Z}.$$

(iii) If $x^1(0) = x^2(0) = 1, x^3(0) = -1$, then the trajectory of the motion has the parametric equations

$$x^1(t) = \frac{1}{1+t},\ x^2(t) = \frac{1}{1+t},\ x^3(t) = -\frac{1}{1+t},\ t \neq -1.$$

(iv) If at least two of the values $x^1(0), x^2(0), x^3(0)$ are different from zero, then either the particle moves to infinity in a finite time (in the future) or it comes from infinity in a finite time (in the past).

Indeed, from $x^1\dot{x}^1 = x^2\dot{x}^2 = x^3\dot{x}^3$ it follows that

$${x^1}^2 - c_1 = {x^2}^2 - c_2 = {x^3}^2 - c_3.$$

Suppose $c_1 \geq c_2 \geq c_3 = 0$. Hence ${x^3}^2 \leq {x^2}^2 \leq {x^1}^2$ and

$${x^3}^2 = {x^1}^2 - c_1 = {x^2}^2 - c_2;\ c_1, c_2 \geq 0;\ \dot{x}^3 = \pm\sqrt{({x^3}^2 + c_1)({x^3}^2 + c_2)}.$$

For simplicity we suppose $x^3(0) \geq 0$ and $\dot{x}^3 = \sqrt{({x^3}^2 + c_1)({x^3}^2 + c_2)}$. It follows that

$$t(x^3) = \int_{x^3(0)}^{x^3} \frac{du}{\sqrt{(u^2 + c_1)(u^2 + c_2)}}$$

and

$$\lim_{x^3 \to \infty} \int_{x^3(0)}^{x^3} \frac{du}{\sqrt{(u^2 + c_1)(u^2 + c_2)}}$$

is finite (since the integral is convergent).

Example 5.6.2. Let $M = \mathbb{R}^n = \{(x^1, ..., x^n)\}$ be the phase space with canonical Nambu bracket of order n and let the first $n-1$ elementary symmetric polynomials of n variables $x^1, ..., x^n$ to be the Hamiltonians:

$$H_1 = s_1, H_2 = s_2, ..., H_{n-1} = s_{n-1},$$

where $(x - x^1) \cdots (x - x^n) = x^n - s_1 x^{n-1} \cdots \pm s_n$. The Nambu-Hamilton equations of motion

$$\dot{x}^i = \{H_1, ..., H_{n-1}, x^i\}$$

can be written as

$$\dot{x}^i = \frac{d^n f}{dx^{i^n}},$$

where $f(x^1, ..., x^n) = \prod_{1 \leq i < j \leq n} (x^i - x^j)$, which might suggest a generalization of a gradient flow.

5.7 Self-evaluation problems

Problem 45. *Show that any Nambu flow conserves the volume.*

Hint The divergence of the vector field $v^i = \varepsilon^{ijk} \partial_j H \partial_k G$ vanishes.

Problem 46. *Show that any change of variables* $x^{i'} = x^{i'}(x)$, $i = 1, 2, 3$, *which conserves the volume in the phase space, i.e.,* $\det \left(\frac{\partial x^{i'}}{\partial x^i} \right) = 1$, *is a symmetry of the Nambu system.*

Problem 47. *Write the Lorenz system*

$$\dot{x} = \sigma(y - x),\ \dot{y} = x(r - z) - y,\ \dot{z} = xy - bz$$

in the Stokes form

$$\dot{\vec{r}} = \nabla H \times \nabla G + \nabla D.$$

Hint Stokes potentials: $H = \frac{1}{2}(y^2 + (z - r)^2)$, $G = \sigma z - \frac{1}{2}x^2$, $D = -\frac{1}{2}(\sigma x^2 + y^2 + bz^2)$.

Problem 48. *Write the Röessler system*

$$\dot{x} = -x - y,\ \dot{y} = x + ay,\ \dot{=} b + z(x - c)$$

in the Stokes form

$$\dot{\vec{r}} = \nabla H \times \nabla G + \nabla D.$$

Hint Stokes potentials: $H = \frac{1}{2}(x + b(1+z))^2 + \frac{1}{2}(y + z + \frac{1}{2}z^2 - b^2)^2$, $G = b \operatorname{arctg} \frac{y+z+\frac{1}{2}z^2 - b^2}{1+b(1+z)}$, $D = -\frac{1}{2}(ay^2 + (x - c)z^2)$.

Problem 49. *Arrange the Euler dynamical system*

$$\dot{x} = ayz,\ \dot{y} = bzx,\ \dot{z} = cxy$$

as Nambu dynamical system.

Problem 50. *Show that the set of all solenoidal vector fields of class C^∞ on $D \subset \mathbb{R}^n$ is a Lie algebra.*

Problem 51. *Study which of the following vector fields are potential fields and, if so, find the potential.*

1) $X = (\sin xy + xy\cos xy)\, i + x^2 \cos xy\, j$;
2) $X = 3y^4z^2\, i + 4x^3z^2\, j - 3x^2y^2\, k$;
3) $X = (y^2\cos x + z^2)\, i - (4 - 2y\sin x)\, j + (2xz + 1)\, k$.

Problem 52. *Determine the harmonic polynomials of two variables.*

Hint The polynomial solutions of the Laplace equation are called *harmonic polynomials*. A basis $\{P_n(x, y), Q_n(x, y)\}$ of the vector space of all the harmonic polynomials of two variables may be defined by $(x + iy)^n = P_n(x, y) + iQ_n(x, y)$, $i^2 = -1$.

Problem 53. *Show that any harmonic function of two variables can be expanded about a regular point into a series of harmonic polynomials having the form*

$$f(x, y) = \sum_{n=0}^{\infty} [a_n P_n(x, y) + b_n Q_n(x, y)].$$

Problem 54. *Determine the harmonic vector fields on $\mathbb{R}^2$, the potentials of which are harmonic polynomials on $\mathbb{R}^2$.*

Bibliography

[1] V. I. Arnold, *Dynamical Systems 3* , Springer-Verlag, Berlin, 1988.

[2] M. Axenides, *Non Hamiltonian chaos from Nambu dynamics of surfaces*, arXiv:1109.0470v1 [nlin.CD] 2 Sep 2011.

[3] R. Chatterjee, *Dynamical symmetries and Nambu Mechanics*, arXiv:hep-th/9501141v2 10 Feb 1995.

[4] S. Codriansky, G. Pereira, C. G. Bernardo, *Nambu-gradient dynamical systems*, Acta Cientifica Venezolana, 53 (2002), 71-73.

[5] S. Lynch, *Dynamical Systems with Applications using Mathematica*, Boston, 2007.

[6] N. Mukunda, E. C. G. Sudarshan, *Relation between Nambu and Hamiltonian Dynamics*, Center for Theoretical Studies, Indian Institute of Science, Bangalore, India, 1975.

[7] L. Takhtajan, *On foundation of the Generalized Nambu Mechanics*, arXiv:hep-th/9301111v1 26 Jan 1993.

[8] C. Udriste, *Field Lines* (in Romanian), Technical Publishing House, Bucharest, 1988.

[9] C. Udriste, *Geometric Dynamics*, Mathematics and Its Applications, 513, Kluwer Academic Publishers, Dordrecht, Boston, London, 2000.

[10] C. Udriste, L. Matei, *Lagrange-Hamilton Theories* (in Romanian), Monographs and Textbooks 8, Geometry Balkan Press, Bucharest, 2008.

[11] C. Udriste, I. Tevy, *Geometric dynamics on Riemannian manifolds*, Mathematics (mdpi), 8, 79 (2020), pp. 1-14.

[12] I. Vaisman, *Lectures on the Geometry of Poisson Manifolds*, Birkhäuser, Berlin, 1994.

Chapter 6

GEOMETRIC DYNAMICS AND WIND THEORY

Motto:

"A sensible man is not the one who always talks
About his wisdom or the one who boasts
That he knows Plato by heart, nor is he wise,
But the one who keeps quiet humbly, bowing."

George Coşbuc - *The Philosophers and the Ploughmen*

In this Chapter, we present some applications selected from science and engineering that refer to the geometric dynamics and wind theory. These are usually formulated by means of dynamical systems that translate physical laws of evolution into mathematical concepts. The winds introduced and described by us are: the biochemical wind, the electric wind, the double pendulum wind, the ABC wind and the economic wind. Suggestions for further reading: [1]-[17].

6.1 Geometric Dynamics generated by a flow and a Riemannian metric

Let $X = (X^1, ..., X^n)$ be a C^∞ vector field on the Riemannian manifold $(\mathbb{R}^n, \delta_{ij})$. The vector field determines the autonomous flow

(1) $$\dot{x}^i(t) = X^i(x(t)), i = \overline{1, n}.$$

The classic extension of the differential system (1) is obtained by derivation along a solution. Explicitly,

$$\ddot{x}^i(t) = \frac{\partial X^i}{\partial x^j}(x(t))\,\dot{x}^j(t),$$

with Einstein convention for summation over the index j. This last differential system has all the solutions of the system (1), but it also has other solutions. Indeed, taking the primitive we obtain $\dot{x}^i(t) = X^i(x(t)) + C^i$, $i = \overline{1,n}$.

We intend to modify the classic extension into an Euler-Lagrange extension. For this we decompose the right hand side into the form (summed over the index j)

$$\ddot{x}^i(t) = \left(\frac{\partial X^i}{\partial x^j} - \frac{\partial X^j}{\partial x^i}\right)(x(t))\,\dot{x}^j(t) + \frac{\partial X^j}{\partial x^i}(x(t))\,\dot{x}^j(t).$$

Replacing $\dot{x}^j(t) = X^j(x(t))$ in the last term, we find the extension

$$\ddot{x}^i(t) = \left(\frac{\partial X^i}{\partial x^j} - \frac{\partial X^j}{\partial x^i}\right)(x(t))\,\dot{x}^j(t) + \frac{\partial f}{\partial x^i}(x(t)), \tag{2}$$

where $f(x) = \frac{1}{2}\,\delta_{ij}X^i(x)X^j(x)$ is the kinetic energy density associated to the vector field X via the Riemannian metric δ_{ij}. The extension (2) turns out to be a system of Euler-Lagrange ODEs. Really, if we use the *least squares Lagrangian*

$$L = \frac{1}{2}\,\delta_{ij}\,(\dot{x}^i - X^i(x))(\dot{x}^j - X^j(x)),$$

then the Euler-Lagrange equations

$$\frac{\partial L}{\partial x^i} - \frac{d}{dt}\frac{\partial L}{\partial \dot{x}^i} = 0$$

are just the equations (2). The second order differential system (2) describes two equivalent notions: (i) a *geodesic motion in a gyroscopic field of forces*, name given by Marsden in a review of the book [12]; (ii) *geometric dynamics*, named given by us (see [4], [11]-[17]). Indeed, on the right side we recognize a gyroscopic force of components

$$\left(\frac{\partial X^i}{\partial x^j} - \frac{\partial X^j}{\partial x^i}\right)\dot{x}^j$$

(which produces zero mechanical work) and a conservative force ∇f (their sum being said also a gyroscopic field of forces).

In summary, from the primordial elements such as a flow and a Riemannian metric (gravitational potentials, masses etc) a geodesic motion appears in a gyroscopic field of forces.

The autonomous least squares Lagrangian L determines the Hamiltonian

$$H = \frac{1}{2}\,\delta_{ij}\,(\dot{x}^i - X^i(x))(\dot{x}^j + X^j(x)),$$

which is conserved on the trajectories.

If we use the Riemannian manifold $(\mathbb{R}^n, g_{ij}(x))$, then the least squares Lagrangian and the Hamiltonian associated to geometric dynamics are

$$L = \frac{1}{2}\,g_{ij}(x)\,(\dot{x}^i - X^i(x))(\dot{x}^j - X^j(x)),$$

$$H = \frac{1}{2}\,g_{ij}(x)\,(\dot{x}^i - X^i(x))(\dot{x}^j + X^j(x)).$$

In this case the movement evolves on the Riemannian manifold $(\mathbb{R}^n, g_{ij}(x))$.

Remark 6.1.1. *In our sense, any geometric dynamics is strongly dependent on the Riemannian manifold* $(M, g_{ij}(x))$. *One of the best selection of the Riemannian manifold adapted to geometric dynamics is by considering the constant curvature: (1) curvature* 0, *Euclidean manifold* $(\mathbb{R}^n, g_{ij}(x) = \delta_{ij})$; *(2) curvature* -1, *hyperbolic manifold* $\left(H^n, g_{ij}(x) = \frac{\delta_{ij}}{x^{n\,2}}\right)$; *(3) curvature* 1, *sphere* $S^n = \{x = (x^1, ..., x^n, x^{n+1}) \in \mathbb{R}^{n+1} \mid \|x\|^2 = 1\}$, *with the metric induced by the Euclidean metric on* $\mathbb{R}^{n+1}$.

6.2 Wind theory and geometric dynamics

Wind refers to the dynamic behaviour in $\mathbb{R}^3$ of the air under conditions of atmospheric disturbance and geometry of the space. Mathematically, the wind is modelled by a velocity vector field (Euler specification)

$$\vec{v}(x^1, x^2, x^3, t),\ (x^1, x^2, x^3) \in \mathbb{R}^3,\ t \in \mathbb{R},$$

and four perturbing forces that apply to the air particles: (1) *inertial force* (represented by total acceleration) $\frac{D\vec{v}}{Dt} = \frac{\partial \vec{v}}{\partial t} + (\vec{v}\,\nabla)\,\vec{v}$; (2)

the pressure gradient force $-\frac{1}{\rho}\operatorname{grad} p$; (3) *the deviating force due to the Earth's rotation* (Coriolis effect, centrifugal force) $\vec{v} \times \vec{F_1}$; (4) *the shearing stresses* (frictional force) produced by the relative motion of the layers above and below, $\vec{F_2}$.

As a result, the evolution PDE of wind appears (similar to Navier Stokes PDE for fluid movement)

$$\underbrace{\frac{D\vec{v}}{Dt}}_{(1)} = \underbrace{-\frac{1}{\rho}\operatorname{grad} p}_{(2)} + \underbrace{\vec{v} \times \vec{F_1}}_{(3)} + \underbrace{\vec{F_2}}_{(4)},$$

with the unknown vector function $\vec{v}(x^1, x^2, x^3, t)$.

6.2.1 Classification of winds

The winds are classified according to the relative importance of the four forces:

(i) *Geostrophic wind.* The movement is stationary (independent of time t). Only forces (2) and (3) remain. This wind is due to the horizontal pressure gradient, the trajectories of the movement being *isobars.*

(ii) *Gradient wind.* The first force is approximated by $\frac{v^2}{R}$, where R is the curvature radius of isobars. The partial acceleration $\frac{\partial \vec{v}}{\partial t}$ and the fourth force are omitted. It remains that the pressure gradient is a quadratic function of velocity.

(iii) *Antitriptic wind.* The forces (2) and (4) are dominant. The force (4) represents friction at the ground and is therefore a force in the opposite direction of the motion, and the wind is towards low pressure.

(iv) *Ageostrophic wind.* It is separated from the geostrophic wind by a frictionless movement. The term (4) is dominant and expresses convection or drags at the Earth's surface.

Because of writing the PDE on the Euclidean space, in the previous explanations the geometry of the space (Riemannian metric, gravitational and/or magnetic potentials) does not seem to be involved. In our opinion, the geometry of the space in which the wind takes place is important and cannot be ignored.

6.2.2 Solitonic wind

Sometimes the behaviour of a wind is like a solitary wave. In other words, the previous PDE exhibits special solutions, known as solitons, which are stable and do not disperse with time. Furthermore there are solutions with more than one soliton which can move towards each other, interact and then emerge at the same speed with no change in shape (but with a time "lag" or "speed up").

Definition 6.2.1. *Denote $\vec{x} = (x^1, x^2, x^3)$, and introduce the wave vector $\vec{k} = (a, b, c)$, the angular frequency ω, the imaginary unit i, and the (possibly complex) amplitude vector $\vec{A}$. A solution of the form $\vec{v}(\vec{x}, t) = \vec{A}\exp(i(\langle\vec{k}, \vec{x}\rangle - \omega t))$ for the wind PDE is called plane wave.*

An interesting case of the plane wave is given in the following

Definition 6.2.2. *A solitary wave solution or soliton $u(x, t)$, $x \in \mathbb{R}^n$, $t \in \mathbb{R}$ is a solution of an evolution PDE that satisfies the following three properties: (i) retains its shape (initial profile) for all time, i.e., $u(x, t) \approx u(x - \alpha t, 0)$, $\alpha \in \mathbb{R}^n$; (ii) is localized (asymptotically constant at $\pm\infty$ or obeys periodicity conditions imposed on the original equation); (iii) can pass through other solitons and retains size and shape.*

Suppose that the previous wind PDE has a soliton solution

$$\vec{v}(\vec{x}, t) = \vec{\varphi}(\xi),\ \xi = \langle\vec{k}, \vec{x}\rangle - \omega t.$$

Then the previous PDE is reduced to the ODE

$$(-\omega + \langle\vec{k}, \vec{\varphi}(\xi)\rangle)\vec{\varphi}\,'(\xi) = -\frac{1}{\rho}\operatorname{grad} p + \vec{\varphi}(\xi) \times \vec{F}_1 + \vec{F}_2.$$

6.2.3 Geometric dynamics as wind

Geometric dynamics in the Riemannian manifold $(\mathbb{R}^3, \delta_{ij})$ is similar to wind theory, except that instead of the velocity vector field $\vec{v}(x^1, x^2, x^3, t)$ on $\mathbb{R}^3 \times \mathbb{R}$ we use the tangent vector field (velocity, Lagrangian specification) $\vec{v}(t) = \vec{v}(x^1(t), x^2(t), x^3(t), t)$, $\vec{v}(t) = \dot{x}^1(t)\vec{i} + \dot{x}^2(t)\vec{j} + \dot{x}^3(t)\vec{k}$, along an arbitrary trajectory $x^1 = x^1(t), x^2 = x^2(t), x^3 = x^3(t), t \in I \subset \mathbb{R}$. Then the term $(\vec{v}\,\nabla)\,\vec{v}$ is missing from the total acceleration.

On $\mathbb{R}^3$, let us consider the ODE system describing the geometric dynamics (geodesic motion in a gyroscopic field of forces)

$$\ddot{x}^i(t) = \left(\frac{\partial X^i}{\partial x^j} - \frac{\partial X^j}{\partial x^i}\right)(x(t))\,\dot{x}^j(t) + \frac{\partial f}{\partial x^i}(x(t)),\ f = \frac{1}{2}\delta_{ij}X^iX^j.$$

Being on $\mathbb{R}^3$, this second order differential system can be written like a wind system

$$\ddot{x}(t) = \dot{x}(t) \times \operatorname{rot} X(x(t)) + \nabla f(x(t)).$$

Let us now explain how the geostrophic wind produced by the previous flow and the Euclidian metric appears. Suppose the vector field $\operatorname{rot} X$ is collinear to $\operatorname{grad} \varphi$, i.e.,

$$\frac{\partial X^i}{\partial x^j} - \frac{\partial X^j}{\partial x^i} = \lambda(x)\frac{\partial \varphi}{\partial x^k},$$

where $\{i, j, k\}$ is a cyclic permutation of $\{1, 2, 3\}$. In this case the geostrofic wind can be written as the differential system

$$\ddot{x} = \operatorname{grad} f + \lambda \operatorname{grad} \varphi \times \dot{x}.$$

Now let us comment the existence of solution (λ, φ) for the equation $\operatorname{rot} X = \lambda \operatorname{grad} \varphi$ (biscalarity condition). The biscalarity condition is locally equivalent to $\langle \operatorname{rot} X, \operatorname{rot}(\operatorname{rot} X)\rangle = 0$. Since this condition is not identically satisfied, the scalar fields λ and φ do not always exist.

If λ and φ are functionally dependent, then it can be proved that X is a potential field; indeed $\lambda(\varphi) \operatorname{grad} \varphi = \operatorname{grad} \psi$ is equivalent to

$$(x) = \int_{x_0}^{x} \lambda(\varphi) d\varphi.$$

Applying the div operator, and taking into account that $\operatorname{div}(\operatorname{rot} X) = 0$, we find the necessary condition $\langle \nabla\lambda, X\rangle + \lambda\Delta\varphi = 0$. Applying the operator rot, and taking into account that $\operatorname{rot}(\operatorname{rot} X) = \operatorname{grad}(\operatorname{div} X) - \Delta X$, we find the necessary condition $\operatorname{grad}(\operatorname{div} X) - \Delta X = \nabla\lambda \times X$.

In the following, we will show in details how certain elementary flows, such as biochemical, electric, double pendulum, Lorenz and surplus demand, generate winds.

6.3 Biochemical geometric dynamics and wind

Negative feedback Some biochemical processes with negative feedback can be modelled by the Goodwin dynamical system [9]

$$\dot{x}(t) = \frac{1}{1+z^n(t)} - ax(t), \dot{y}(t) = x(t) - by(t), \dot{z}(t) = y(t) - cz(t),$$

$(x, y, z) \in \mathbb{R}^3_+$, where a, b, c are strictly positive parameters, n is a fixed natural number and x, y, z are functions of concentration.

First, let us show that significant solutions for biology have images in a closed bounded region in the concentration space (the first quadrant) $x \geq 0, y \geq 0, z \geq 0$. Particularly, we shall describe an invariant set with respect to the flow attached to the Goodwin differential system.

Since x, y, z are concentration functions, the initial conditions $x \geq 0, y \geq 0, z \geq 0$ are perfectly suited for biological applications. These initial conditions and the positiveness of derivatives $\dot{x}, \dot{y}, \dot{z}$ for $x = y = z = 0$ show that $x(t) \geq 0, y(t) \geq 0, z(t) \geq 0, \forall t > 0$. The inequality $z \geq 0$ implies $\dot{x}(t) = \frac{1}{1+z^n(t)} - ax(t) \leq 1 - ax(t)$ and consequently: (1) if $x(0) \leq \frac{1}{a}$, then $x(t) \leq \frac{1}{a}, \forall t > 0$; if $x(0) > \frac{1}{a}$, then the derivative $\dot{x}(t)$ is negative as long as $x(t) > \frac{1}{a}$ and hence $x(t)$ will decrease, becoming after a while at least $\frac{1}{a}$. In conclusion $0 \leq x(t) \leq \max\left\{x(0), \frac{1}{a}\right\}, \forall t > 0$. The inequality $x(t) < \frac{1}{a}$ implies $\dot{y}(t) = x - by \leq \frac{1}{a} - by$ and arguments similar to the preceding ones shows that $0 \leq y(t) \leq \max\left\{y(0), \frac{1}{ab}\right\}, \forall t > 0$. Analogously, $0 \leq z(t) \leq \max\left\{z(0), \frac{1}{abc}\right\}, \forall t > 0$. Implicitly we have shown that the parallelepiped

$$\bar{D} = \left\{(x, y, z) \,|\, 0 \leq x \leq \frac{1}{a},\ 0 \leq y \leq \frac{1}{ab},\ 0 \leq z \leq \frac{1}{abc}\right\} \subset \mathbb{R}^3_+$$

is an invariant set (the orbits determined by the initial points in $\bar{D}$ are included in $\bar{D}$, and the orbits entering in $\bar{D}$ stay in $\bar{D}$). This result is completed by the remark that the divergence of the Goodwin vector field $X = (X^1, X^2, X^3)$ is strictly negative, i.e., $\text{div}X = -a - b - c$, and hence the flow generated by X decreases the volume.

Let us show that the Goodwin dynamical system admits a single equilibrium point belonging to the parallelepiped $\bar{D}$. For this we consider the algebraic system

$$\frac{1}{1+z^n} - ax = 0,\ x - by = 0,\ y - cz = 0;\ x, y, z \geq 0$$

whose solutions generate the equilibrium points. Eliminating x and y, we obtain the equation $\frac{1}{1+z^n} = abcz$, which has a single strictly positive root z_0. There exists a single equilibrium point $(x_0 = bcz_0, y_0 = cz_0, z_0)$. Since $\psi(z) = 1 - abcz(1+z^n)$ satisfies $\psi(0) = 1 > 0, \psi\left(\frac{1}{abc}\right) < 0$ it follows that

$$0 < z_0 < \frac{1}{abc},\ 0 < y_0 < \frac{1}{ab},\ 0 < x_0 < \frac{1}{a}.$$

Let us analyse the stability of the equilibrium point (x_0, y_0, z_0) using the linear approximation with the Jacobi matrix

$$A = \begin{bmatrix} \frac{\partial X^i}{\partial x} & \frac{\partial X^i}{\partial y} & \frac{\partial X^i}{\partial z} \end{bmatrix} (x_0, y_0, z_0),$$

i.e.,

$$\begin{bmatrix} -a & 0 & \frac{-nz_0^{n-1}}{1+z_0^n} \\ 1 & -b & 0 \\ 0 & 1 & -c \end{bmatrix}$$

The characteristic equation of the matrix A is $(\lambda + a)(\lambda + b)(\lambda + c) + \frac{nz_0^{n-1}}{1+z_0^n} = 0$. This equation certainly has a real root λ_1 (as an equation of degree 3), and necessarily $\lambda_1 < 0$ (since all the coefficients of the equation are strictly positive). As the preceding equation can be written

$$\lambda^3 + (a+b+c)\lambda^2 + (ab+ac+bc)\lambda + abc + \frac{nz_0^{n-1}}{1+z_0^n} = 0,$$

the Hurwitz criterion shows that all the roots have strictly negative real parts if and only if

$$(a+b+c)(ab+ac+bc) - abc - \frac{nz_0^{n-1}}{1+z_0^n} > 0.$$

In this hypothesis the equilibrium point (x_0, y_0, z_0) is asymptotically stable.

We compute

$$\operatorname{rot} X = \left(1, \frac{nz^{n-1}}{(1+z^n)^2}, 1\right),$$

$$f(x, y, z) = \frac{1}{2}\delta_{ij}X^i X^j = \frac{1}{2}\left(\left(\frac{1}{1+z^n} - ax\right)^2 + (x - by)^2 + (y - cz)^2\right).$$

Let us consider $x = (x^1, x^2, x^3) = (x, y, z)$. Then the *biochemical wind* is the second order dynamical system

$$\ddot{x}(t) = \dot{x}(t) \times \operatorname{rot} X(x(t)) + \nabla f(x(t)).$$

Since $(X, \operatorname{rot} X) \neq 0$, this wind is not a geostrophic wind.

Positive feedback There exist biochemical processes with positive feedback described mathematically by the dynamical system [9]

$$\dot{x}(t) = \frac{1 + z^n(t)}{k + z^n(t)} - ax(t), \dot{y}(t) = x(t) - by(t), \dot{z}(t) = y(t) - cz(t),$$

$(x, y, z) \in \mathbb{R}^3_+$, where a, b, c are strictly positive parameters, k is a supraunitary real parameter, n is a fixed natural number and x, y, z are functions of concentration.

The function $\varphi : [0, \infty) \to \mathbb{R}, \varphi(z) = \frac{1+z^n}{k+z^n}$ admits the derivative $\varphi'(z) = \frac{nz^{n-1}(k-1)}{(k+z^n)^2} > 0$ and hence it is strictly increasing. Since

$$\varphi(0) = \frac{1}{k}, \lim_{z\to\infty} \varphi(z) = 1, \lim_{z\to 0} \varphi'(z) = 0, \lim_{z\to\infty} \varphi'(z) = 0,$$

the graph of the function φ has the shape in Fig. 1.

> plot$((z^7 + 1)/(1.5 + z^7), z = 0..10)$

The inequalities $\dot{x} = \frac{1+z^n}{k+z^n} - ax \leq 1 - ax, x \geq 0, z \geq 0$ together with arguments similar to those used for "negative feedback" lead to $0 \leq x(t) \leq \max\left\{x(0), \frac{1}{a}\right\}, \forall t > 0$. Then we deduce

$$0 \leq y(t) \leq \max\left\{y(0), \frac{1}{ab}\right\}, \ 0 \leq z(t) \leq \max\left\{z(0), \frac{1}{abc}\right\}, \forall t > 0.$$

Consequently the parallelepiped

$$\bar{D} = \left\{(x, y, z)\,|\, 0 \leq x \leq \frac{1}{a}, \ 0 \leq y \leq \frac{1}{ab}, \ 0 \leq z \leq \frac{1}{abc}\right\} \subset \mathbb{R}^3_+$$

is an invariant set (the orbits determined by initial points in $\bar{D}$ are included in $\bar{D}$, and the orbits entering in $\bar{D}$ stay in $\bar{D}$).

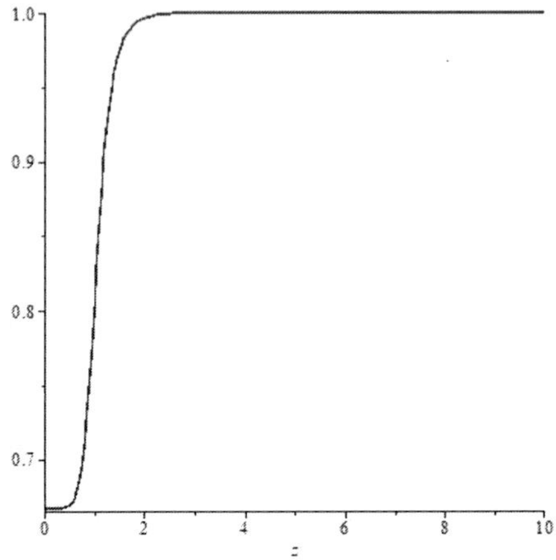

Figure 6.1: Graph of function φ

The divergence of the vector field $X = (X^1, X^2, X^3)$, $X^1(x, y, z) = \frac{1+z^n}{k+z^n} - ax$, $X^2(x, y, z) = x - by$, $X^3(x, y, z) = y - cz$ is strictly negative, i.e., $\operatorname{div} X = -a - b - c$. Therefore the flow generated by X decreases the volumes.

This dynamical system has at least one equilibrium point in $\bar{D}$. Indeed, from $\frac{1+z^n}{k+z^n} - ax = 0, x - by = 0, y - cz = 0$ we deduce the equation $\frac{1+z^n}{k+z^n} = abcz$ which has at least one positive root z_0. We obtain the equilibrium point $(x_0 = bcz_0, y_0 = cz_0, z_0)$ which belongs to $\bar{D}$. The matrix of the corresponding linear approximation is

$$\begin{bmatrix} -a & 0 & \varphi'(z_0) \\ 1 & -b & 0 \\ 0 & 1 & -c \end{bmatrix}$$

and has the characteristic equation

$$\lambda^3 + (a + b + c)\lambda^2 + (ab + ac + bc)\lambda + abc - \varphi'(z_0) = 0.$$

The Hurwitz criterion shows that all the roots of this equation have strictly negative real parts if and only if

$$(a + b + c)(ab + ac + bc) > abc - \varphi'(z_0) > 0.$$

In these conditions the equilibrium point (x_0, y_0, z_0) is asymptotically stable.

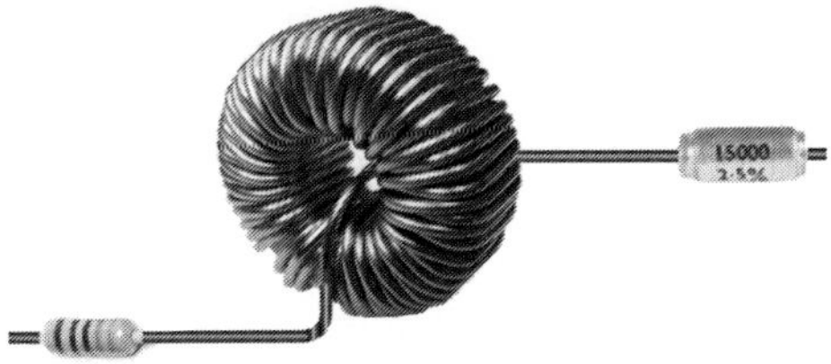

Figure 6.2: RLC electrical circuit

We find

$$\operatorname{rot} X = \left(1, \frac{nz^{n-1}(k-1)}{(1+z^n)^2}, 1\right),$$

$$f(x,y,z) = \frac{1}{2}\delta_{ij}X^iX^j = \frac{1}{2}\left(\left(\frac{1+z^n}{k+z^n} - ax\right)^2 + (x-by)^2 + (y-cz)^2\right).$$

Let $x = (x^1, x^2, x^3) = (x, y, z)$. Then the *biochemical wind* is the second order dynamical system

$$\ddot{x}(t) = \dot{x}(t) \times \operatorname{rot} X(x(t)) + \nabla f(x(t)).$$

Since $(X, \operatorname{rot} X) \neq 0$, this wind is not a geostrophic wind.

6.4 Electric geometric dynamics and wind

An *RLC* circuit is an electrical circuit consisting of a resistor (R), an inductor (L), and a capacitor (C), connected in series or in parallel (Fig. 2).

Denote by x the electric charge of the capacitor, with $\dot{x}$ the current in the circuit, and by $f(x, \dot{x})$ a nonlinear function satisfying: $f(0,0) = 0$ and its Taylor expansion around $(0,0)$ is of order greater than 2. Then the function $x(t)$ satisfies the second order differential equation

$$L\ddot{x}(t) + R\dot{x}(t) + \frac{1}{C}\,x(t) + f(x(t), \dot{x}(t)) = 0,$$

where L, R, C are strictly positive numbers. This second order differential equation is equivalent to the first order dynamical system

$$\dot{x}(t) = y(t),\ \dot{y}(t) = -\frac{1}{LC}\,x(t) - \frac{R}{L}\,y(t) - \frac{1}{L}f(x(t), y(t)),$$

having $(0,0)$ as equilibrium point. The matrix of linear approximation

$$\dot{x}(t) = y(t), \ \dot{y}(t) = -\frac{1}{LC}\, x(t) - \frac{R}{L}\, y(t),$$

has the characteristic equation $\lambda^2 + \frac{R}{L}\,\lambda + \frac{1}{LC} = 0$. If $R^2 \neq \frac{4L}{C}$, then the equilibrium point $(0,0)$ is asymptotically stable. This result is a consequence of physical reasons: if the Ohm resistance is positive, then the current ceases inevitably by disappearing in time.

The previous theory is 2-dimensional. For the calculation of the rotor we need a 3-dimensional space and a 3-dimensional vector field. For consistency we introduce $x = (x^1, x^2, x^3 = 0)$, $X = (X^1, X^2, X^3 = 0)$, where

$$X^1(x,y) = y, \ X^2(x,y) = -\frac{1}{LC}\, x(t) - \frac{R}{L}\, y(t) - \frac{1}{L} f(x(t), y(t)),$$

and

$$f(x,y) = \frac{1}{2}\left(y^2 + \left(\frac{1}{LC}\, x + \frac{R}{L}\, y + \frac{1}{L} f(x,y)\right)^2\right).$$

We find

$$\mathrm{rot}X = \left(0, 0, -\frac{1}{LC} - \frac{1}{L} f_x - 1\right), \ \mathrm{div}X = -\frac{R}{L} - \frac{1}{L} f_y.$$

Consequently: (i) if $R + f_y < 0$, then the flow increases area; (ii) if $R + f_y > 0$, then the flow decreases area; (iii) if $R + f_y = 0$, then the flow conserves area.

Since $(X, \mathrm{rot}X) = 0$, the vector field X is biscalar. On the other hand, the vector field $\mathrm{rot}X$ is also biscalar. Indeed, the vector field

$$\mathrm{rot}X = \left(0, 0, -\frac{1}{LC} - \frac{1}{L} f_x - 1\right)$$

can be written in the form $\mathrm{grad}\,\varphi$, where

$$\varphi(x,y,z) = -\left(\frac{1}{LC} + \frac{1}{L} f_x + 1\right) z.$$

This means that the electric wind

$$\ddot{x}(t) = \dot{x}(t) \times \mathrm{grad}\,\varphi(x(t)) + \nabla f(x(t))$$

is a geostrophic wind.

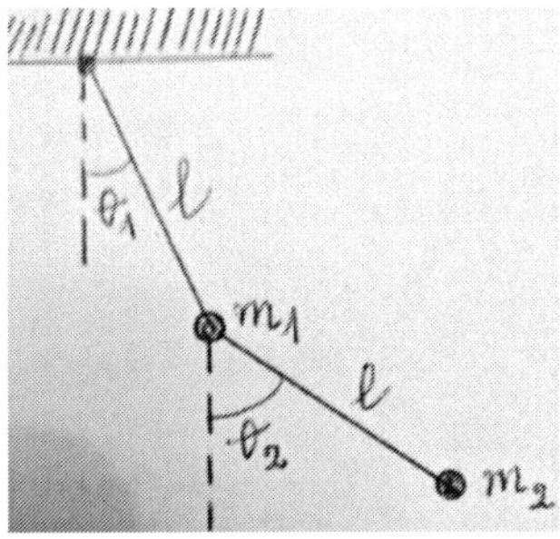

Figure 6.3: Double pendulum

6.5 Double pendulum geometric dynamics and wind

In physics and mathematics, a double pendulum (Fig. 3) is a pendulum with another pendulum attached to its end, and is a simple physical system that exhibits rich dynamic behaviour with a strong sensitivity to initial conditions. The motion of a double pendulum is governed by a set of coupled ordinary differential equations and is chaotic.

The generalized coordinates defining the configuration of the system are the angles between each limb and the vertical, θ_1 and θ_2. If the origin of the Cartesian coordinate system is taken to be at the point of suspension of the first pendulum, then the center of mass of this pendulum is the point $x_1 = \frac{l}{2}\sin\theta_1, y_1 = -\frac{l}{2}\cos\theta_1$ and and the center of mass of the second pendulum is the point

$$x_2 = l\left(\sin\theta_1 + \tfrac{1}{2}\sin\theta_2\right), \; y_2 = -l\left(\cos\theta_1 + \tfrac{1}{2}\cos\theta_2\right).$$

The Lagrangian $L =$ kinetic energy $-$ potential energy is

$$L = \tfrac{1}{2}m\left(v_1^2 + v_2^2\right) + \tfrac{1}{2}I\left(\dot{\theta}_1^2 + \dot{\theta}_2^2\right) - mg\left(y_1 + y_2\right)$$

$$= \tfrac{1}{2}m\left(\dot{x}_1^2 + \dot{y}_1^2 + \dot{x}_2^2 + \dot{y}_2^2\right) + \tfrac{1}{2}I\left(\dot{\theta}_1^2 + \dot{\theta}_2^2\right) - mg\left(y_1 + y_2\right).$$

The first term is the linear kinetic energy of the center of mass of the bodies and the second term is the rotational kinetic energy around the center of mass of each rod. The last term is the potential energy of the bodies in a uniform gravitational field.

Substituting the coordinates above and rearranging the terms, we obtain

$$L = \tfrac{1}{6}ml^2\left(\dot{\theta}_2^2 + 4\dot{\theta}_1^2 + 3\dot{\theta}_1\dot{\theta}_2\cos(\theta_1 - \theta_2)\right) + \tfrac{1}{2}mgl\left(3\cos\theta_1 + \cos\theta_2\right).$$

Then the expressions of momenta are

$$p_{\theta_1} = \frac{\partial L}{\partial \dot{\theta}_1} = \tfrac{1}{6}ml^2\left(8\dot{\theta}_1 + 3\dot{\theta}_2\cos(\theta_1 - \theta_2)\right)$$
$$p_{\theta_2} = \frac{\partial L}{\partial \dot{\theta}_2} = \tfrac{1}{6}ml^2\left(2\dot{\theta}_2 + 3\dot{\theta}_1\cos(\theta_1 - \theta_2)\right).$$

These last expressions may be inverted to get

$$\dot{\theta}_1 = \frac{6}{ml^2}\frac{2p_{\theta_1} - 3\cos(\theta_1 - \theta_2)p_{\theta_2}}{16 - 9\cos^2(\theta_1 - \theta_2)}$$
$$\dot{\theta}_2 = \frac{6}{ml^2}\frac{8p_{\theta_2} - 3\cos(\theta_1 - \theta_2)p_{\theta_1}}{16 - 9\cos^2(\theta_1 - \theta_2)}.$$

The remaining equations of motion are written as

$$\dot{p}_{\theta_1} = \frac{\partial L}{\partial \theta_1} = -\tfrac{1}{2}ml^2\left(\dot{\theta}_1\dot{\theta}_2\sin(\theta_1 - \theta_2) + 3\frac{g}{l}\sin\theta_1\right)$$
$$\dot{p}_{\theta_2} = \frac{\partial L}{\partial \theta_2} = -\tfrac{1}{2}ml^2\left(-\dot{\theta}_1\dot{\theta}_2\sin(\theta_1 - \theta_2) + \frac{g}{l}\sin\theta_2\right).$$

These last four differential equations are explicit formulae for the time evolution of the system given its current state. It is not possible to go further and integrate these equations analytically, to get formulae for θ_1 and θ_2 as functions of time.

The least squares Lagrangian is

$$2\mathcal{L} = \left(p_{\theta_1} - \frac{\partial L}{\partial \dot{\theta}_1}\right)^2 + \left(p_{\theta_2} - \frac{\partial L}{\partial \dot{\theta}_2}\right)^2 + \left(\dot{p}_{\theta_1} - \frac{\partial L}{\partial \theta_1}\right)^2 + \left(\dot{p}_{\theta_2} - \frac{\partial L}{\partial \theta_2}\right)^2.$$

The Euler-Lagrange equations of $\mathcal{L}$ defines the double pendulum wind.

6.5.1 Variational integrator

The method of solving an Euler-Lagrange system by discretization (as example, Runge-Kutta method) is not appropriate because we accumulate too many errors, and the numerical solution is far from the exact solution. The way to go is to discretize the action, write the discrete Euler-Lagrange equations (variational integrator) and then numerically solve them.

Let us consider $x \in \mathbb{R}^n$ and let $\gamma = \{(t, x) \,|\, x = x(t), t_0 \le t \le t_1\}$ be a curve in the space $\mathbb{R} \times \mathbb{R}^n$. Let Γ be the set of all curves γ joining the points (t_0, x_0) and (t_1, x_1). A differentiable function

$$L : \mathbb{R} \times \mathbb{R}^n \times \mathbb{R}^n \to \mathbb{R}, \ (t, x, \dot{x}) \to L(t, x, \dot{x})$$

is called Lagrangian density energy.

Theorem 6.5.1. *A C^2 curve γ is an extremal of the functional (action)*

$$E(\gamma) = \int_{t_0}^{t_1} L(t, x(t), \dot{x}(t)) \, dt$$

if and only if it is a solution of Euler-Lagrange ODEs system

$$\frac{\partial L}{\partial x^k} - \frac{d}{dt}\frac{\partial L}{\partial \dot{x}^k} = 0, \ k = 1, ..., n.$$

The Euler-Lagrange equations are second order differential equations in the unknown $x(t)$. In fact they are n equations, and the solutions depend on $2n$ arbitrary constants. For fixing one solution we use $2n$ boundary conditions $x(t_0) = x_0, x(t_1) = x_1$. The Euler-Lagrange ODEs do not depend on the choice of a system of coordinates, since the Euler-Lagrange operator behaves like a tensor.

The Lagrangian L determines the Hamiltonian

$$H(t, x, \dot{x}) = \dot{x}^i \, \frac{\partial L}{\partial \dot{x}^i}(t, x, \dot{x}) - L(t, x, \dot{x}).$$

The discretization of the Lagrangian L can be made by using the midpoint rule which consists in the substitution of t by h, of the point x with $\frac{1}{2}(x_{k+1} + x_k)$, and of the velocity $\dot{x}$ by $\frac{1}{h}(x_{k+1} - x_k)$, where h is the time step. One obtains the discrete Lagrangian

$$L_d : \mathbb{R} \times \mathbb{R}^n \times \mathbb{R}^n \to \mathbb{R}, \ L_d(h, u, v) = L\left(h, \frac{v + u}{2}, \frac{v - u}{2}\right).$$

This determines the discrete action

$$S : \mathbb{R} \times (\mathbb{R}^n)^{N+1} \to \mathbb{R}, \ S(h, x_0, x_1, ..., x_N) = \sum_{k=0}^{N-1} L(h, x_k, x_{k+1}),$$

$x_k \in \mathbb{R}^n$. The discrete variational principle characterizes the sequence $(x_0, x_1, ..., x_N) \in (\mathbb{R}^n)^{N+1}$ for which the action S is stationary, for any family $x_k(\epsilon) \in \mathbb{R}^n$, $\epsilon \in I \subset \mathbb{R}$, $0 \in I$, with $x_k(0) = x_k$, and x_0, x_N fixed points. Using the variation of the first order we find

Theorem 6.5.2. *The sequence* (x_k), $k = 0, 1, ..., N$, *is stationary for the action* S *if and only if it is generated by the discrete Euler-Lagrange equations (variational integrator)*

$$\frac{\partial L_d}{\partial x_k^i}(h, x_{k-1}, x_k) + \frac{\partial L_d}{\partial x_k^i}(h, x_k, x_{k+1}) = 0,\ i = 1, ..., n; k = 1, ..., N-1.$$

6.6 ABC geometric dynamics and wind

We use the Riemannian manifold $(\mathbb{R}^3, \delta_{ij})$. One example of a fluid velocity that contains exponential stretching and hence unstability is the ABC flow

$$\begin{aligned}\dot{x}^1(t) &= A\sin x^3(t) + C\cos x^2(t),\\ \dot{x}^2(t) &= B\sin x^1(t) + A\cos x^3(t),\\ \dot{x}^3(t) &= C\sin x^2(t) + B\cos x^1(t).\end{aligned}$$

The name of this flow is given after the three mathematicians Arnold, Beltrami, and Childress, who have significantly contributed to our understanding and appreciation of classes of chaotic flows of which the present one is an example. For nonzero values of the constants A, B, C the preceding system is not globally integrable. The topology of the flow lines is very complicated and can only be investigated numerically to reveal regions of chaotic behaviour. The ABC flow conserves the volumes since the ABC field is solenoidal.

The ABC geometric dynamics (Euler-Lagrange prolongation) is described by

$$\ddot{x}^i(t) = \frac{\partial f}{\partial x^i} + \left(\frac{\partial X^i}{\partial x^j} - \frac{\partial X^j}{\partial x^i}\right)\dot{x}^j(t), i, j = 1, 2, 3,$$

where

$$2f = A+B+C+2AC\sin x^3\cos x^2+2BA\sin x^1\cos x^3+2CB\sin x^2\cos x^1.$$

Since $\operatorname{rot} X = X$, the ABC geometric dynamics is equivalent to the ABC wind

$$\ddot{x}(t) = \nabla f(x(t)) + \dot{x}(t) \times X(x(t)).$$

Using

$$2L = \delta_{ij}(\dot{x}^i - X^i(x))(\dot{x}^j - X^j(x)), 2H = \delta_{ij}(\dot{x}^i - X^i(x))(\dot{x}^j + X^j(x))$$

$$g_{ij} = (H + f)\delta_{ij}, N^i_j = \Gamma^i_{jk}y^k - F^i_j,$$

$$F^i_j = \delta^{ih}F_{jh}, F_{ij} = \frac{\partial X^i}{\partial x^j} - \frac{\partial X^j}{\partial x^i},$$

the solutions of the previous differential system are horizontal pregeodesics of the Riemann-Jacobi-Lagrange manifold $\left(\mathbb{R}^3 \setminus \mathcal{E}, g_{ij}, N^i_j\right)$, where $\mathcal{E}$ is the set of equilibrium points.

6.7 Economic geometric dynamics and wind

Let us explain the uni-temporal geometric dynamics produced by the vector field "excess demand".

The vector field E of "excess demand" has the components

$$E_1(p_1, p_2, p_3) = -\frac{p_2}{p_1 + p_2} + \frac{p_3}{p_3 + p_1}$$

$$E_2(p_1, p_2, p_3) = -\frac{p_3}{p_2 + p_3} + \frac{p_1}{p_1 + p_2}$$

$$E_3(p_1, p_2, p_3) = -\frac{p_1}{p_3 + p_1} + \frac{p_2}{p_2 + p_3}.$$

On $(\mathbb{R}^3_+, \delta_{ij})$, we consider the excess demand dynamical system

$$\frac{dp_i}{dt} = E_i(p_1, p_2, p_3),\ i = 1, 2, 3.$$

The equilibrium set is the semi-line $\{(p_1, p_1, p_1) \mid p_1 > 0\}$. The general theory shows that a non-isolated equilibrium point can be stable but not asymptotically stable.

The positive homogeneity of order zero of each component E_i means if all the prices are multiplied by the same constant, the surplus demand does not vary. The Walras law $\sum_{i=1}^3 p_i E_i = 0$ means that the function

$$f_1(p_1, p_2, p_3) = \frac{1}{2}(p_1^2 + p_2^2 + p_3^2)$$

is a first integral. The constant level sets of this first integral are spheres (compact surfaces, invariant with respect to the excess demand flow).

The excess demand vector field satisfies also the Tzitzeica law

$$\sum p_1 p_2 E_3 = 0 \text{ (summation after cyclic permutations of } \{1, 2, 3\}).$$

That is why, the second first integral of this flow is $f_2(p_1, p_2, p_3) = p_1 p_2 p_3$, i.e., the excess demand flow conserves the Tzitzeica surfaces. Economically, this means the volume of prices is constant.

In conclusion, excess demand flow consists in orbits of the form

$$p_1^2 + p_2^2 + p_3^2 = C_1, \; p_1 p_2 p_3 = C_2.$$

The previous dynamical system and the Euclidean metric determine the least squares Lagrangian

$$L = \frac{1}{2}\delta^{ij}\left(\frac{dp_i}{dt} - E_i\right)\left(\frac{dp_j}{dt} - E_j\right).$$

The Euler-Lagrange equations follow

$$\frac{d^2 p_i}{dt^2} = \delta^{jk}\left(\frac{\partial E_i}{\partial p_j} - \frac{\partial E_j}{\partial p_i}\right)\frac{dp_k}{dt} + \frac{\partial f}{\partial p_i},$$

where

$$f = \frac{1}{2}\delta^{ij} E_i E_j$$

is the density of energy produced by the vector field E and the Euclidean metric. Equivalent, it appears the *single-time geometric dynamics* with the associated Hamiltonian

$$H = \frac{1}{2}\delta^{ij}\left(\frac{dp_i}{dt} - E_i\right)\left(\frac{dp_j}{dt} + E_j\right) = \frac{1}{2}\delta^{ij}\frac{dp_i}{dt}\frac{dp_j}{dt} - f.$$

Since $\operatorname{rot} E$ can be written in the form $\operatorname{grad} \varphi$, where

$$\varphi(p_1, p_2, p_3) = \frac{p_3}{p_1 + p_2} + \frac{p_1}{p_2 + p_3},$$

we find the economic wind generated by the excess demand vector field E.

6.8 Self-evaluation problems

Problem 55. *Transform the Euler flow*

$$\dot{x} = ayz,\ \dot{y} = bzx,\ \dot{z} = cxy$$

in a wind using the Euclidean metric.

Problem 56. *Transform the Goodwin flow*

$$\dot{x} = \frac{1}{1+z^n} - ax,\ \dot{y} = x - by,\ \dot{z} = y - cz$$

in a wind using the Euclidean metric.

Problem 57. *The flow $\dot{x} = y,\ \dot{y} = z,\ \dot{z} = x$ is included in a gyroscopic motion (wind). Find the gyroscopic force.*

Problem 58. *The Euler PDEs for an ideal, incompressible, homogeneous fluid moving in the region Ω are*

$$\frac{\partial \vec{v}}{\partial t} + (\vec{v}\,\nabla)\vec{v} = -\nabla p,$$

with the constraint $div\,\vec{v} = 0$ and the boundary conditions: $\vec{v}$ is tangent to the boundary $\partial\Omega$. Show that these PDEs have a Hamiltonian structure.

Solution. Let $\frac{\delta F}{\delta \vec{v}}$ be defined by

$$\lim_{\epsilon\to 0} \frac{1}{\epsilon}[F(\vec{v}+\epsilon\delta\vec{v}) - F(\vec{v})] = \int_\Omega \langle \delta\vec{v}, \frac{\delta F}{\delta\vec{v}}\rangle\ d^3x.$$

Introduce the (+) Lie-Poisson bracket, called the ideal fluid bracket, on functions of $\vec{v}$ by

$$\{F,G\}(\vec{v}) = \int_\Omega \langle \vec{v}, \left[\frac{\delta F}{\delta\vec{v}}, \frac{\delta G}{\delta\vec{v}}\right]_L \rangle\ d^3x,$$

where the index L means left Lie algebra bracket. With the energy function chosen to be the kinetic energy,

$$H(\vec{v}) = \frac{1}{2}\int_\Omega \|\vec{v}\|^2 d^3x,$$

one can verify that the Euler PDEs are equivalent to the Poisson bracket equations $\dot{F} = \{F,H\}$.

Problem 59. *Show that any C^1 vector field X admits an orthogonal decomposition of the form $X = \mathbb{P}X + \nabla\varphi$, into a divergence-free part $\mathbb{P}X$ and a gradient part $\nabla\varphi$.*

Problem 60. *For a fixed function $f : \mathbb{R}^3 \to \mathbb{R}$, we define $\{F, K\}_f = \langle \nabla f, \nabla F \times \nabla K\rangle$, where all three functions are smooth. Show that $\{F, K\}_f$ is a Poisson bracket.*

Problem 61 (Continuation). *Show that any divergence-free vector field X on $\mathbb{R}^3$ can be written globally as a "rot" of another vector field and, away from equilibrium points, can locally be written as $X = \nabla f \times \nabla g$, where f and g are smooth real-valued functions on $\mathbb{R}^3$. Assume this (so-called Clebsch-Monge) representation also holds globally. Show that the particles of fluid, which follow trajectories satisfying $\dot{x}(t) = X(x(t))$, are trajectories of a Hamiltonian system with a bracket in the form of the previous problem.*

Problem 62. *The electric vector field E and the magnetic vector field B define the Hamiltonian*

$$H(E, B) = \frac{1}{2}\int \left(\|E(x,t)\|^2 + \|B(x,t)\|^2\right) d^3x.$$

Write the Hamiltonian PDEs.

Problem 63. *Suppose a mean-reverting geometrical Brownian motion (GBM) mean-reverts towards a known weak diurnal cycle. We can model wind strength X by a stochastic differential equation (SDE)*

$$dX = k_X\left(\theta_X(t) + \frac{1}{k_x}\dot{\theta}_X(t) - X(t)\right) dt + \sigma_X X(t) dW,$$

where k_X is the constant local speed of mean-reversion of X towards its local mean $\theta_X(t)$ while σ_X is a constant volatility and dW is the increment of a Wiener variable. The parameter values assumed here are $k_X = 0.1h^{-1}$ and $\sigma_X = 0.2h^{-1/2}$ and the daily seasonal forcing function for wind is

$$\theta_X(t) = \bar{\theta}_X\left(1 + \alpha_X \sin(\gamma(t + \psi_X))\right),$$

where $\bar{\theta}_X = 8ms^{-1}$, $\alpha_X = 0.375$, $\psi_X = 2h$, $\gamma = \frac{\pi}{12}h^{-1}$. Find $X(t)$.

Bibliography

[1] A. Bossavit, *Differential forms and the computation of fields and forces in electromagnetism*, Eur. J. Mech., B, Fluids, 10, no 5(1991), 474-488.

[2] C. Fox, *An Introduction to the Calculus of Variations*, Dover Publications, Inc., New York, 1963.

[3] Q. Huang, *Solving an open sensor exposure problem using variational calculus*, Report Number: WUCSE-2003-1 (2003). All Computer Science and Engineering Research. https://openscholarship.wustl.edu/cse-research/1057.

[4] D. Isvoranu, C. Udriste, *Fluid flow versus Geometric Dynamics*, 5-th Conference on Differential Geometry, August 28-September 2, 2005, Mangalia, Romania; BSG Proceedings 13, pp. 70-82, Geometry Balkan Press, 2006.

[5] M. L. Krasnov, G. I. Makarenko, A. I. Kiselev, *Problems and Exercises in the Calculus of Variations*, Mir Publisheres, Moscow, 1975.

[6] E. Langamann, *Introduction to Variational Calculus*, INTERNET, 2008.

[7] I. N. Popescu, *Gravitation*, Editrice Nagard, Roma, Italy, 1988.

[8] I. B. Russak, *Calculus of Variations*, MA4311, Lectures Notes, Department of Mathematics, Naval Postgraduate School, Monterey, California, 93943, July, 2002.

[9] L. Segel, *Mathematical Models in Molecular and Cellular Biology*, Cambridge, UK, Cambridge University Press, 1980.

[10] F. Suryawan, J. D. Dona, M. Seron, *Methods for trajectory generation in magnetic levitation system under constraints*, 2010 annual report - University of New Castle.

[11] C. Udriste, *Geometric dynamics*, Southeast Asian Bulletin of Mathematics, Springer-Verlag, 24, 1 (2000), 313-322.

[12] C. Udriste, *Geometric Dynamics*, Mathematics and Its Applications, 513, Kluwer Academic Publishers, Dordrecht, Boston, London, 2000.

[13] C. Udriste, M. Ferrara, D. Opris, *Economic Geometric Dynamics*, Monographs and Textbooks 6, Geometry Balkan Press, Bucharest, 2004.

[14] C. Udriste, *Tools of geometric dynamics*, Buletinul Institutului de Geodinamică, Academia Română, 14, 4 (2003), 1-26; Proceedings of the XVIII Workshop on Hadronic Mechanics, honoring the 70-th birthday of Prof. R. M. Santilli, the originator of hadronic mechanics, University of Karlstad, Sweden, June 20-22, 2005; Eds. Valer Dvoeglazov, Tepper L. Gill, Peter Rowland, Erick Trell, Horst E. Wilhelm, Hadronic Press, International Academic Publishers, December 2006, ISBN 1-57485-059-28, pp 1001-1041.

[15] C. Udriste, A. Udriste, *From flows and metrics to dynamics and winds*, Bull. Cal. Math. Soc., 98, 5 (2006), 389-394.

[16] C. Udriste, L. Matei, *Lagrange-Hamilton Theories* (in Romanian), Monographs and Textbooks 8, Geometry Balkan Press, Bucharest, 2008.

[17] C. Udriste, D. Zugravescu, F. Munteanu, *Nonclassical electromagnetic dynamics*, WSEAS Transactions on Mathematics, 7, 1 (2008), 31-39.

Chapter 7

MOTIONS OF CURVES

Motto:

"With cheerful voices of young disposition,
Remembering the old glory,
We are heading ahead in thought towards the sun,
We have the light and the good as our aim -
May long live our country and people!"

George Coşbuc - *The Students' Hymn*

Generally, a motion is the change in the position of an object over time. The motion is mathematically described in terms of displacement, distance, velocity, acceleration, speed, and time. The motion of a body is observed by attaching a frame of reference to an observer and measuring the change in position of the body relative to that frame.

In this Chapter, we analyse the motion of a parameterized curve, in $\mathbb{R}^2$ *and in* $\mathbb{R}^3$*, adding to the parameter that describes the curve a second evolution parameter and specifying a PDE whose solution is the variation. This study is similar to considering a surface and analyzing coordinate curves by Frenet type formulas. Both constant torsion curves and constant curvature curves lead to sine-Gordon PDE of compatibility. Other problems treated are: generating surface, the flow on the tangent vector, the flow on the normal vector (the flow of the fastest descent of the length of a curve) and the flow on the binormal vector. Such problems have direct applications in science and engineering, being related to evolutionary phenomena. Suggestions for further reading: [1]-[8].*

7.1 Generating surfaces

To understand the idea of moving a curve, we start with the surfaces in $\mathbb{R}^3$ generated by straight lines or circles. Indeed, the simplest ways to generate a surface are the movement of a straight line or the rotation of a curve around an axis. A generated surface is usually described in the language of the free vectors or in the implicit Cartesian language. Now we describe it as a solution of a Cauchy problem.

7.1.1 Ruled surfaces

A surface in $\mathbb{R}^3$ described by the vector function (position vector)

$$\vec{r}(u,v) = \vec{\alpha}(u) + v\vec{\beta}(u),\ u \in I,\ v \in \mathbb{R}$$

is called *ruled surface.* Such a surface is generated by the movement of a straight line ($u = c$) which rests on a curve ($v = 0$). Dynamically a ruled surface appears as the solution to the Cauchy problem

$$\partial_v \vec{r}(u,v) = \vec{\beta}(u),\ \vec{r}(u,v)|_{v=0} = \vec{\alpha}(u).$$

The straight lines $u = c$ are called *rulings.* The curve $\vec{\alpha}(u)$ is called *director curve.*

If $\vec{\beta}(u)$ is a non-zero constant vector (represents a direction), then the ruled surface is called a *cylindrical surface.* In this case the straight line $u = c$ evolves through parallelism and rests on the director curve.

If $\vec{\alpha}(u)$ is a constant vector (represents a point), then the ruled surface is called *conical surface.* In this case the straight line $u = c$ evolves passing by a fixed point called *vertex* and leans on a *director curve.*

7.1.2 Surfaces of revolution

A surface in $\mathbb{R}^3$ that can be generating by revolving a curve C around a fixed straight line D is called *surface of revolution.* The curve C is called *generatrix*, and the straight line D is called *axis of revolution.*

A vector field on $\mathbb{R}^3$ of the form $X(x) = Ax$, where A is an antisymmetric matrix, is called *Killing vector field.* If $\vec{a} = (a^1, a^2, a^3)$ and $\vec{r} = (x^1, x^2, x^3)$, then $\vec{a} \times \vec{r}$ is a Killing vector field. Conversely, any Killing vector field on $\mathbb{R}^3$ can be written in the form $\vec{a} \times \vec{r}$, since any 3×3 antisymmetric matrix A is equivalent to a vector $\vec{a}$.

Theorem 7.1.1. *A surface of revolution is a field surface of a Killing vector field of the form $X(x) = Ax$.*

Proof. Let $X(x) = (X^1(x), X^2(x), X^3(x))$ be a vector field of class C^∞. The constant level sets of a function $f : \mathbb{R}^3 \to \mathbb{R}$ of class C^2, which satisfies the partial derivative equation

$$(PDE) \qquad X^1(x)\frac{\partial f}{\partial x^1}(x) + X^2(x)\frac{\partial f}{\partial x^2}(x) + X^3(x)\frac{\partial f}{\partial x^3}(x) = 0,$$

are called *field surfaces*. The characteristic system associated to this PDE is

$$\frac{dx^1}{X^1(x)} = \frac{dx^2}{X^2(x)} = \frac{dx^3}{X^3(x)}.$$

If $g(x)$ and $h(x)$ are two first integrals of this system, then the general solution of the PDE is $f = \Phi(h, g)$, where Φ is an arbitrary function of class C^1.

Let $X(x) = Ax$ be a Killing vector field on $\mathbb{R}^3$. Since the matrix A is antisymmetric, the orbits of the Killing vector field are described by the Cartesian implicit equations $||x||^2 = c_1, ||Ax||^2 = c_2$. It follows that the function $f(x) = \Phi(||x||^2, ||Ax||^2)$, where Φ is an arbitrary function of class C^1, is the general solution of the PDE $\langle \nabla f(x), Ax \rangle = 0$. Therefore the surfaces $f(x) = c$ are surfaces of revolution.

Conversely, if we have a surface of revolution fixed by the vector $\vec{a}$, then $\vec{a} \times \vec{r}$ is a Killing vector field tangent to this surface. □

Remark 7.1.1. *The theory in the following paragraphs includes the movement of thin wires (curves as free rigid bodies) through roto-translations. A large class of physical systems are modeled in terms of the motion of curves in Euclidean space. A remarkable example is the so-called localized induction equation $x_t = x_s \times x_{ss}$ which is a soliton equation used to model the dynamics of a thin vortex filament in an incompressible, inviscid, homogeneous, 3-dimensional fluid.*

7.2 Motion of a curve in space

In mathematical modelling of diverse nonlinear events of the natural and the applied sciences such as vortex filament dynamics, interface movements, robot arm shape control, flame front propagation, image processing, overwrapped DNA, magnetic fluxes, membrane deformation and protein dynamics, the motions of space curves are being used. The evolutions of these nonlinear phenomena are described by the differential equations which characterize the motions of curves as a family.

We start from a curve represented by the position vector $\vec{\beta}: J \to \mathbb{R}^3$, $\vec{\beta} = \vec{\beta}(s)$, of class C^3. Suppose it has unit speed $||\vec{\beta}'|| = 1$, namely, the parameter that describes the curve is the curvilinear abscissa s. We attach the *curvature* $k > 0$, *torsion* τ and the orthonormal moving frame $\{\vec{T}, \vec{N}, \vec{B}\}$, called *Frenet frame*, where $\vec{T}$ is called the *unit tangent vector field*, $\vec{N}$ is the *unit normal vector field* and $\vec{B}$ is the *unit binormal vector field.* Along this curve the *Frenet formulas* or *Frenet ODEs*,

$$\vec{T}' = k\vec{N},\ \vec{N}' = -k\vec{T} + \tau\vec{B},\ \vec{B}' = -\tau\vec{N},$$

hold, where the symbol $'$ means the derivative with respect to the parameter s (curvilinear abscissa).

The least squares Lagrangian

$$2L(\vec{T}, \vec{N}, \vec{B}; \vec{T}', \vec{N}', \vec{B}') = \|\vec{T}' - k\vec{N}\|^2 + \|\vec{N}' + k\vec{T} - \tau\vec{B}\|^2 + \|\vec{B}' + \tau\vec{N}\|^2$$

defines a geometric dynamics and hence a Frenet wind.

7.2.1 The integration of the natural equations of a curve

The general theory of first order ODEs guarantees the existence and uniqueness (up to a rigid motion) for the Frenet ODEs, as soon as the curvature and torsion are given.

Let us show that finding analytically a curve when the curvature and torsion are given is quite another matter. First, we remark that the Frenet system contains three copies of the scalar system

$$X'(s) = k(s)Y(s),\ Y'(s) = -k(s)X(s) + \tau(s)Z(s),\ Z'(s) = -\tau(s)Y(s).$$

The solution (X, Y, Z) must satisfy $X^2 + Y^2 + Z^2 = 1$. This last equation can be decomposed (over the complex numbers) as $(X + iY)(X - iY) = (1 + Z)(1 - Z)$. Defining the complex conjugate functions

$$u = \frac{X + iY}{1 - Z},\ -\frac{1}{v} = \frac{X - iY}{1 - Z},$$

we find the complex parametrization

$$X = \frac{1 - uv}{u - v},\ Y = i\,\frac{1 + uv}{u - v},\ Z = \frac{u + v}{u - v}.$$

The solution of the Frenet system is obtained finding the complex functions $u(s)$ and $v(s)$. Each of $u(s)$ and $v(s)$ is a solution of the Riccati ODE

$$\frac{dw}{ds}(s) = -\frac{i}{2}\ \tau(s) - i\ k(s)\ w(s) + \frac{i}{2}\ \tau(s)\ w(s)^2.$$

This offers an indirect proof of the fact that generally the Frenet ODE system cannot be integrated through quadratures.

A time evolution of the parameterized curve $\vec{\beta}(s)$ is obtained by introducing the time t as the second parameter and imposing a PDE of temporal evolution. We accept that by evolution we obtain a surface (we avoid sliding along the curve, reparametrizations, extensions, etc). Thus, we may assume (without loss of generality) that any variation of a regular curve consists of regular curves.

Let's suppose, we know already the C^2 variation surface

$$\vec{r} = \vec{r}(s,t),\ \vec{r}(s,0) = \vec{\beta}(s),\ s \in J,\ t \in [0,\varepsilon),$$

guided by a given evolution PDE.

Now the Frenet ODEs of the curve $s \to \vec{r}(s,t)$ transcribe as Frenet PDEs

$$(1) \qquad \frac{\partial}{\partial s}\vec{T} = k\vec{N},\ \frac{\partial}{\partial s}\vec{N} = -k\vec{T} + \tau\vec{B},\ \frac{\partial}{\partial s}\vec{B} = -\tau\vec{N}.$$

Analogous, to the temporal evolution $t \to \vec{r}(s,t)$ we attach the orthonormal Frenet moving frame $\{\vec{T},\vec{N},\vec{B}\}$ which satisfies the PDE system

$$(2) \qquad \frac{\partial}{\partial t}\vec{T} = a\vec{N} + b\vec{B},\ \frac{\partial}{\partial t}\vec{N} = -a\vec{T} + c\vec{B},\ \frac{\partial}{\partial t}\vec{B} = -b\vec{T} - c\vec{N}.$$

The PDE system (1) + (2) must be completely integrable, i.e.,

$$\frac{\partial^2}{\partial s\partial t} = \frac{\partial^2}{\partial t\partial s}.$$

It follows the complete integrability conditions (existence of a solution)

$$(CIC) \qquad a_s = k_t + b\tau,\ b_s = kc - \tau a,\ c_s = \tau_t - kb,$$

where the indices mean partial derivatives. In the relations (CIC), we accept implicitly the fact that $\vec{\beta}$ is an inextensible curve, i.e.,

$||\vec{r}_s(s,0)|| = ||\vec{r}_s(s,t)|| = 1$ (the length of the initial curve = the length of deformed curve). The relations (CIC) imply

$$(a^2 + b^2 + c^2)_s = 2(ak_t + c\tau_t).$$

Details can be made only in particular cases.

The PDEs (1) + (2) suggest the least squares Lagrangian

$$2L = \|\vec{T}_s - k\vec{N}\|^2 + \|\vec{N}_s + k\vec{T} - \tau\vec{B}\|^2 + \|\vec{B}_s + \tau\vec{N}\|^2$$

$$+\|\vec{T}_t - a\vec{N} - b\vec{B}\|^2 + \|\vec{N}_s + a\vec{T} - c\vec{B}\|^2 + \|\vec{B}_s + b\vec{T} + c\vec{N}\|^2.$$

The Euler-Lagrange PDEs attached to this least squares Lagrangian refer to a Frenet wind.

7.2.2 A motion of an inextensible curve of constant torsion

Suppose τ = constant (constant torsion). We add the simplifying condition $a = 0$, so that the relationships (2) to be of type (1). Then $b^2+c^2 = d^2(t)$ or in parametric form $b = d(t)\sin\sigma$, $c = d(t)\cos\sigma$. The (CIC) relations are reduced to $\sigma_{st} = -d(t)\tau(s)\sin\sigma$, where $k = \sigma_s$. Particularizing again, $\tau = \tau_0 = \frac{1}{\rho}$ (constant) and $d = -\frac{1}{\rho}$, we find the sine-Gordon equation

$$\sigma_{st} = \frac{1}{\rho^2}\sin\sigma, \tag{3}$$

whose unknown is the angle σ. However, the PDE system (1) + (2), which describes the generated surface, becomes

$$\begin{pmatrix} \vec{T} \\ \vec{N} \\ \vec{B} \end{pmatrix}_s = \begin{pmatrix} 0 & \sigma_s & 0 \\ -\sigma_s & 0 & \frac{1}{\rho} \\ 0 & -\frac{1}{\rho} & 0 \end{pmatrix} \begin{pmatrix} \vec{T} \\ \vec{N} \\ \vec{B} \end{pmatrix}$$

$$\begin{pmatrix} \vec{T} \\ \vec{N} \\ \vec{B} \end{pmatrix}_t = \begin{pmatrix} 0 & 0 & -\frac{1}{\rho}\sin\sigma \\ 0 & 0 & -\frac{1}{\rho}\cos\sigma \\ \frac{1}{\rho}\sin\sigma & \frac{1}{\rho}\cos\sigma & 0 \end{pmatrix} \begin{pmatrix} \vec{T} \\ \vec{N} \\ \vec{B} \end{pmatrix}.$$

The geometric significance is as follows: a constant torsion curve associated with the sine-Gordon PDE describes a pseudo-spherical surface, of curvature $K = -\frac{1}{\rho^2} < 0$, and at each moment t the curve of parameter s is an asymptotic line of this surface.

The evolution PDE of the pseudo-sphere (surface described above) is

$$\vec{r}_t = \cos\sigma\,\vec{T} - \sin\sigma\,\vec{N},$$

(Theme!).

Remark 7.2.1. *(i) Let K be the Gauss curvature of a surface in $\mathbb{R}^3$. The classification of the surfaces according to Gaussian curvature is: (1) for $K = 1$, it is a sphere, (2) for $K = 0$, it is a plane, (3) for $K = -1$, it is a pseudo-sphere.*

(ii) A parametric representation of the pseudo-sphere is

$$x(u,v) = \rho\ sech\left(\frac{u}{\rho}\right)\cos\frac{v}{\rho}\,,$$

$$y(u,v) = \rho\ sech\left(\frac{u}{\rho}\right)\sin\frac{v}{\rho}\,,$$

$$z(u,v) = \rho\ \left(\frac{u}{\rho} + \alpha - tanh\left(\frac{u}{\rho} + \alpha\right)\right),$$

where u = const represents the parallels, and v = const represents the meridians (surface of revolution).

7.2.3 A motion of an inextensible curve of constant curvature

Suppose k = constant (constant curvature). We add the simplifying condition $c = 0$. The (CIC) relations imply

$$a = -\frac{1}{\rho}\cos\sigma,\ b = \frac{1}{\rho}\sin\sigma,\ k = \frac{1}{\rho},\ \tau = \sigma_s,$$

where the angle σ is a solution of sine-Gordon PDE (3). In this case, the PDEs (1)+(2), which define the generated surface, are reduced to PDEs

$$\begin{pmatrix}\vec{T}\\ \vec{N}\\ \vec{B}\end{pmatrix}_s = \begin{pmatrix}0 & \frac{1}{\rho} & 0\\ -\frac{1}{\rho} & 0 & \sigma_s\\ 0 & -\sigma_s & 0\end{pmatrix}\begin{pmatrix}\vec{T}\\ \vec{N}\\ \vec{B}\end{pmatrix}$$

$$\begin{pmatrix}\vec{T}\\ \vec{N}\\ \vec{B}\end{pmatrix}_t = \begin{pmatrix}0 & -\frac{1}{\rho}\cos\sigma & \frac{1}{\rho}\sin\sigma\\ \frac{1}{\rho}\cos\sigma & 0 & 0\\ -\frac{1}{\rho}\sin\sigma & 0 & 0\end{pmatrix}\begin{pmatrix}\vec{T}\\ \vec{N}\\ \vec{B}\end{pmatrix}.$$

The curve used here is no longer located on a pseudo-spherical surface, nor can we decide what the surface is.

7.3 Motion of curve along the tangent vector

We start with a spatial C^3 curve $\vec{\beta}: I \to \mathbb{R}^3$, $s \to \vec{\beta}(s)$, with unit tangent vector field $\vec{T}$, unit normal vector field $\vec{N}$, unit binormal vector field $\vec{B}$, orthonormal Frenet moving frame $\{\vec{T}, \vec{N}, \vec{B}\}$, curvature $k > 0$ and torsion τ.

A time-differentiable evolution of the curve $\vec{\beta}$ described as a variation is

$$\vec{r} = \vec{r}(s,t),\ \vec{r}(s,0) = \vec{\beta}(s),\ s \in I,\ t \in [0, \varepsilon),$$

guided by a given evolution PDE. On the other hand we recall that if s is the curvilinear abscissa of the curve $\vec{\beta}(s) = \vec{r}(s,0)$, then $k(s)\vec{N}(s) = \vec{\beta}''(s) = \partial_{ss}\vec{r}(s,0)$.

Firstly, we accept that the variation satisfies the Cauchy problem (PDE + initial condition)

$$\partial_t \vec{r}(s,t) = \vec{T}(s),\ \vec{r}(s,0) = \vec{\beta}(s)$$

(motion along the tangent vector of the initial curve). It follows the ruled surface

$$\vec{r}(s,t) = \vec{\beta}(s) + t\,\vec{T}(s),\ t > 0,$$

which is called *tangent surface* of the curve β or *tangent developable.* A surface that can be rolled out, without stretching or shrinking onto a plane is called *developable.*

The general definition of the tangent flow (Cauchy problem)

$$(EDP) \qquad \partial_t \vec{r}(s,t) = \vec{T}(s,t),\ \vec{r}(s,0) = \vec{\beta}(s)$$

can be detailed only on concrete cases concerning $\vec{T}(s,t)$. Obviously we also need the compatibility condition $\vec{r}_{ts} = \vec{r}_{st}$.

7.4 Motion of curve along normal vector

This propagation model is known as Euclidean geometric heat flow, as well as mean curvature motion.

For simplification, we start with a plane curve represented by the position vector $\vec{\alpha}: I \to \mathbb{R}^2$ of class C^2, with unit tangent vector field $\vec{T}$, unit normal vector filed $\vec{N}$, the orthonormal Frenet moving frame $\{\vec{T}, \vec{N}\}$, oriented such that the curvature k to be positive (see the

Frenet formulas in space; the versor $\vec{N}$ indicates the bending sense of the curve). A differentiable variation in time is described using the conditions

$$\vec{r} = \vec{r}(s,t),\ \vec{r}(s,0) = \vec{\alpha}(s),\ s \in I,\ t \in [0,\varepsilon).$$

We accept that this variation satisfies the Cauchy problem

$$(PDE) \qquad \partial_t \vec{r}(s,t) = k(s,t)\vec{N}(s,t),\ \ \vec{r}(s,0) = \vec{\alpha}(s)$$

(motion along normal direction, scaled by curvature) and we take into account the conditions of complete integrability $\vec{r}_{ts} = \vec{r}_{st}$. On the other hand, since s is the curvilinear abscissa of the curve $\vec{\alpha}(s) = \vec{r}(s,0)$, we have

$$\vec{T}(s) = (x'(s), y'(s)),\ \ k(s)\vec{N}(s) = \vec{\alpha}''(s) = \partial_{ss}\vec{r}(s,0).$$

Let's show that the previous variation is actually a flow that shortens length of the curves.

Theorem 7.4.1. *Let's consider the length functional (transcribed like integral with parameter)* $\ell(t) = \displaystyle\int_0^1 ||\vec{r}_s(s,t)||\, ds$, *where* $\vec{r}(s,t)$ *is an arbitrary variation of a curve* $\vec{\alpha}(s) = \vec{r}(s,0)$ *which satisfies the boundary value problem*

$$\vec{r}_t(s,t) = \vec{V}(s,t),\ \vec{V}(0,t) = 0,\ \vec{V}(L,t) = 0,$$

with $\vec{V}$ *normalized by the condition* $\displaystyle\int_0^L ||\vec{V}(s,t)||^2\, ds = 1$, *where* $L = \ell(0)$, *and* s *is the curvilinear abscissa of the curve* α. *Then the rate of variation* $\frac{d\ell}{dt}$ *of the length is maximal if* $\vec{V}$ *is collinear to* $k\vec{N}$.

Proof. First, we calculate the partial derivative

$$\partial_t\ ||\vec{r}_s|| \ = \ \partial_t < \vec{r}_s, \vec{r}_s >^{\frac{1}{2}} \ = \ \frac{1}{||\vec{r}_s||} < \vec{r}_s, \partial_t \vec{r}_s > \ = \ \frac{1}{||\vec{r}_s||} < \vec{r}_s, \partial_s \partial_t \vec{r} >$$

$$= < \vec{r}_s, \frac{\partial_s \vec{V}}{||\vec{r}_s||} > = < \vec{r}_s, \partial_s \vec{V} > .$$

It follows

$$\frac{d\ell}{dt} = \int_0^1 \partial_t ||\vec{r}_s||\, ds = \int_0^L < \vec{r}_s, \partial_s \vec{V} > \frac{1}{||r_s||} ds = \int_0^L < \partial_s \vec{r}, \partial_s \vec{V} > ds$$

$$=< \partial_s \vec{r}, \vec{V} > |_0^L - \int_0^L < \partial_{ss}\vec{r}, \vec{V} > ds = \int_0^L < -k\vec{N}, \vec{V} > ds$$

$$\leq \left(\int_0^L k^2 ds \right)^{\frac{1}{2}} \left(\int_0^L ||\vec{V}||^2 ds \right)^{\frac{1}{2}},$$

with equality if and only if the vectors $\vec{N}$ and $\vec{V}$ are collinear. □

Example 7.4.2 (Shrinking circle). Let us discuss about the *circle that tightens.* Let be the circle $\vec{\alpha}(u) = r_0(\cos u, \sin u)$ submersed in the plane disk (circular crown) $\vec{r}(u,t) = R(t)(\cos u, \sin u)$. In this case $\vec{N} = -\frac{1}{R}\vec{r}$, $k = \frac{1}{R}$ and the evolution equation is reduced to $\frac{dR}{dt} = -\frac{1}{R}$. It follows $R = (r_0^2 - 2t)^{\frac{1}{2}}$ and $\vec{r}(u,t) = (r_0^2 - 2t)^{\frac{1}{2}}(\cos u, \sin u)$. The circle tends to a point (origin) when $t \to \frac{1}{2}r_0^2$.

In mathematics, the curve-shortening flow is a process that modifies a smooth curve in the Euclidean plane by moving its points perpendicularly to the curve at a speed proportional to the curvature. The curve-shortening flow is an example of a geometric flow, and it is the one-dimensional case of the mean curvature flow. Other names for the same process include the *Euclidean shortening flow, geometric heat flow*, and *arc-length evolution.*

7.4.1 Existence of the PDE solution that describes the movement along normal

The motion along normal reduces to a partial derivative equation (PDE). We use the Theorem of existence and uniqueness of parabolic equations, to show that this PDE has solutions for t small enough. We start with the curve $\vec{\alpha}(u) = (x(u), y(u))$ of class C^2, we pass to a differentiable variation

$$\vec{r}(u,t) = (x(u,t), y(u,t)) \ \vec{r}(u,0) = \vec{\alpha}(u), \ u \in I, \ t \in [0, \varepsilon)$$

and transcribe the normal evolution PDE in the system form

$$\begin{pmatrix} x \\ y \end{pmatrix}_t = \frac{1}{(x_u^2 + y_u^2)^2} \begin{pmatrix} y_u^2 & -x_u y_u \\ -x_u y_u & x_u^2 \end{pmatrix} \begin{pmatrix} x_{uu} \\ y_{uu} \end{pmatrix},$$

with initial condition $\vec{r}(u,0) = \vec{\alpha}(u)$.

Theorem 7.4.3. *If the immersion $\vec{\alpha}(u)$ is smooth for $u \in I$, then the initial PDE problem has a smooth solution $\vec{r}(u,t)$, $u \in I$, $t \in [0, \varepsilon)$.*

Proof. The previous PDEs system could be parabolic if the matrix

$$\begin{pmatrix} y_u \\ -x_u \end{pmatrix} \begin{pmatrix} y_u & -x_u \end{pmatrix} = \begin{pmatrix} y_u^2 & -x_u y_u \\ -x_u y_u & x_u^2 \end{pmatrix}$$

would be positive definite. Obviously it is not so because this matrix is degenerate. In fact, "the parabolic character of this system of PDEs is hidden by the poor parameterization" of the phenomena. However, how do we show that the solution exists?

Suppose that we know a solution $\vec{r}(u,t) = (x(u,t), y(u,t))$ and we denote with $\varphi_t : I \to I, t \in [0, \varepsilon)$ a family of diffeomorphisms, from the curve $\vec{\alpha}$ to itself, indexed by t. We introduce the composition $\vec{q}(u,t) = \vec{r}(\varphi(u,t), t) = (\bar{x}(u,t), \bar{y}(u,t))$, which is in fact a reparametrization.

The evolution PDE definition

$$\partial_t \varphi(u,t) = V(\varphi(u,t), t)$$

leads to

$$\vec{q}_t(u,t) = \partial_t \vec{r}(\varphi(u,t), t) + \vec{r}_u \; \partial_t \varphi(u,t) = -k\vec{N} + \vec{r}_u V$$

$$= \frac{1}{(x_u^2 + y_u^2)^2} \begin{pmatrix} y_u \\ -x_u \end{pmatrix} \begin{pmatrix} y_u & -x_u \end{pmatrix} \begin{pmatrix} x_{uu} \\ y_{uu} \end{pmatrix} + \begin{pmatrix} x_u \\ y_u \end{pmatrix} V.$$

If we add the choice

$$V = \frac{1}{(x_u^2 + y_u^2)^2} \begin{pmatrix} x_u & y_u \end{pmatrix} \begin{pmatrix} x_{uu} \\ y_{uu} \end{pmatrix},$$

we attend a strong parabolic system. To prove this, we first calculate

$$\begin{pmatrix} \bar{x} \\ \bar{y} \end{pmatrix}_t$$

$$= \frac{1}{(x_u^2 + y_u^2)^2} \left\{ \begin{pmatrix} y_u \\ -x_u \end{pmatrix} \begin{pmatrix} y_u & -x_u \end{pmatrix} + \begin{pmatrix} x_u \\ y_u \end{pmatrix} \begin{pmatrix} x_u & y_u \end{pmatrix} \right\} \begin{pmatrix} x_{uu} \\ y_{uu} \end{pmatrix}$$

$$= \frac{1}{x_u^2 + y_u^2} \begin{pmatrix} 1 & 0 \\ 0 & 1 \end{pmatrix} \begin{pmatrix} x_{uu} \\ y_{uu} \end{pmatrix}.$$

Then we express the derivatives x_u, x_{uu}, y_u, y_{uu} in function of $\bar{x}_u, \bar{x}_{uu}$, $\bar{y}_u, \bar{y}_{uu}$, using the derivation of the composite functions. We get a strong parabolic PDE system because the matrix that multiplies the vector $(\bar{x}_{uu}, \bar{y}_{uu})$ is positive definite.

By the existence and uniqueness Theorem for a diffusion system (heat equation), there is a solution $(\bar{x}, \bar{y})$ on the temporal interval $[0, \varepsilon)$ and this solution is unique. By means of $(\bar{x}, \bar{y})$ we build the vector field V and than we have the solution $\varphi(u, t)$. We introduce $\vec{r}(u,t) = \vec{q}(\varphi^{-1}(u,t), t)$ and we show that this vector satisfies the starting PDE. In this way, although the initial PDEs system is not parabolic, by the previous reparameterization it has a solution on the interval $[0, \varepsilon)$. The bound ε can be estimated as

$$\varepsilon = \sup\{\tau \mid \vec{r}(u,t) \,\text{exists and is smooth for}\; t \in [0, \tau)\}.$$

□

7.5 Korteweg-de Vries (KdV) equation

We resume the ideas in the Section 3. Let $\{\vec{T}, \vec{N}, \vec{B}\}_{(s,t)}$ be the Frenet moving frame. We accept that the PDE

$$\vec{r}_t(s,t) = \lambda(s,t)\vec{T}(s,t) + \mu(s,t)\vec{N}(s,t) + \nu(s,t)\vec{B}(s,t),$$

describes the temporal evolution of the initial curve. The complete integrability condition $\vec{r}_{ts} = \vec{r}_{st}$ implies

$$(\lambda_s - \mu k)\vec{T} + (\lambda k + \mu - \nu\tau)\vec{N} + (\mu\tau + \nu_s)\vec{B} = a\vec{N} + b\vec{B},$$

i.e.,

$$\lambda_s - \mu k = 0,\; \lambda k + \mu - \nu\tau = a,\; \mu\tau + \nu_s = b.$$

The temporal evolution of curvature k and torsion τ can be expressed by the components (λ, μ, ν) of the partial velocity $\vec{r}_t$. Substituting a and b in the conditions of complete integrability (CIC) in Section 2, we find

(3) $$k_t = (\lambda k + \mu_s - \nu\tau)_s - (\mu\tau + \nu_s)\tau,$$

(4) $$\tau_t = \gamma_s + (\mu\tau + \nu_s)k,$$

where

$$c = \frac{1}{k}\left[(\mu\tau + \nu_s)_s + \tau(\lambda k + \mu_s - \nu\tau)\right].$$

We particularize by choosing an inextensible curve with zero torsion (plane curve) moving under the condition $\mu = -k_s$. It follows

$\lambda = -\frac{k^2}{2} + C(t)$, where $C(t)$ is an arbitrary function. Choosing $C(t) = 0$, we deduce

$$a = -k_{ss} - \frac{k^3}{2},\ b = \nu_s,\ c = \frac{\nu_{ss}}{k},$$

$$\lambda = -\frac{k^2}{2},\ \mu = -k_s,\ \nu = \text{arbitrary},$$

while the relationship (3) shows that the curvature k evolves according to the KdV PDE

$$k_t + k_{sss} + \frac{3}{2}\, k^2 k_s = 0$$

(the unknown is the curvature k). The constraint (4) imposes a condition upon the binormal component of the partial velocity $\vec{r}_t$, namely,

$$\left(\frac{\nu_{ss}}{k}\right)_s = -k\nu_s.$$

7.6 Motion of curve along the binormal vector

We consider an inextensible spatial C^3 curve $\vec{\beta} : I \to \mathbb{R}^3$ having the curvature $k > 0$ and the torsion τ. The unit tangent field $\vec{T}$, the unit normal field $\vec{N}$ and the unit binormal field $\vec{B}$ determines the orthonormal Frenet moving frame $\{\vec{T}, \vec{N}, \vec{B}\}$. An evolution of the curve in time is described by a differentiable variation surface

$$\vec{r} = \vec{r}(s,t),\ \vec{r}(s,0) = \vec{\beta}(s),\ s \in I,\ t \in [0, \varepsilon).$$

The PDE of the binormal flow can be conceived as

$$\partial_t \vec{r}(s,t) = \partial_s \vec{r}(s,t) \times \partial_{ss} \vec{r}(s,t).$$

The parameterization condition by the arc-length $||\partial_s \vec{r}||^2 = 1$ is compatible with the previous equation because

$$\partial_t ||\partial_s \vec{r}||^2 = 2(\partial_s \vec{r}, \partial_{st} \vec{r})) = 2(\partial_s \vec{r}, \partial_s \vec{r} \times \partial_{sss} \vec{r}) = 0.$$

Particularly, all the curves evolving according to this rule have the same length.

7.6.1 Nonlinear Schrödinger equation (NSE)

Usually, in geometrical terms, for the binormal flow an equivalent PDE

$$(PDE) \qquad \partial_t \vec{r}(s,t) = k(s,t)\vec{B}(s,t),\ \vec{r}(s,0) = \vec{\beta}(s)$$

is used (binormal vector field scaled by curvature). It can be shown that this evolution leads to a solitonic PDE which incorporates PDEs of Dym type

$$u_t = u^3 u_{xxx},$$

of Camassa-Holm type

$$u_t + 2au_x - u_{xxt} + 3uu_x = 2u_x u_{xx} + uu_{xxx},\ a = \text{const},$$

(met in fluid dynamics) or Schrödinger type (quantum states of a physical system).

The connection between motion of space or plane curves and integrable equations has drawn wide interest in the past and many results have been obtained. One pioneering work is due to Hasimoto who showed that the nonlinear Schrödinger PDE describes the motion of an isolated non-stretching thin vortex filament.

Studying the approximation of self-induced motion of a thin isolated vortex filament travelling without stretching in an incompressible fluid, Hasimoto (1972) [4] took as a model the movement of a curve along the binormal vector, where $\vec{r} = \vec{r}(s,t)$ is the vectorial equation of the vortex filament in movement. Using the notations in Section 2, we find $a = -k\tau, b = k_s$, with the compatibility condition $ck = k_{ss} - k\tau^2$ and with time evolution PDEs

$$(5) \qquad k_t = -2k_s\tau - k\tau_s,\ \tau_t = \left(-\tau^2 + \frac{k_{ss}}{k} + \frac{k^2}{2}\right)_s$$

of curvature k and torsion τ. By Hasimoto transformation

$$q = ke^{i\sigma},\ \sigma = \int_{s_0}^{s} \tau(\xi,t)d\xi,$$

the previous PDEs become

$$q_t = \left[k_t + ik\left(-\tau^2 + \frac{k_{ss}}{k} + \frac{k^2}{2} - A(t)\right)\right]e^{i\sigma}$$

$$q|q|^2 = k^3 e^{i\sigma}, q_{ss} = (k_{ss} + 2ik_s\tau + ik\tau_s - k\tau^2)e^{i\sigma},$$

where

$$A(t) = \left(-\tau^2 + \frac{k_{ss}}{k} + \frac{k^2}{2}\right)|_{s=s_0}.$$

We transcribe it in the form

$$q_t = i(q_{ss} + \frac{1}{2}q|q|^2 - A(t)q).$$

A new function change,

$$Q = q\exp\left(i\int_0^t A(\xi)d\xi\right),$$

leads to Schrödinger PDE

$$i\,Q_t + Q_{ss} + \frac{1}{2}\,|Q|^2\,Q = 0.$$

If $A(t) = 0$, then the version with q of the Schrödinger PDE admits the linear reprezentation (PDEs system)

$$\begin{pmatrix} \vec{T} \\ \vec{N} \\ \vec{B} \end{pmatrix}_s = \begin{pmatrix} 0 & k & 0 \\ -k & 0 & \tau \\ 0 & -\tau & 0 \end{pmatrix} \begin{pmatrix} \vec{T} \\ \vec{N} \\ \vec{B} \end{pmatrix}$$

$$\begin{pmatrix} \vec{T} \\ \vec{N} \\ \vec{B} \end{pmatrix}_t = \begin{pmatrix} 0 & -k\tau & -k_s \\ k\tau & 0 & \frac{k_{ss}}{k} - \tau^2 \\ -k_s & -\frac{k_{ss}}{k} + \tau^2 & 0 \end{pmatrix} \begin{pmatrix} \vec{T} \\ \vec{N} \\ \vec{B} \end{pmatrix}.$$

The surfaces generated by the binormal motion of a curve are called *Hasimoto surfaces* or *Schrödinger surfaces.*

7.6.2 Solitonic Hasimoto surface

We resume the system of partial derivative equations (5). Trying to find a solitary wave to propagate at constant speed ω along the vortex filament, Hasimoto imposed the condition $\lim_{s\to\infty} k(s,t) = 0$ and a solution of type

$$k = k(\xi),\ \xi = s - \omega t;\ \tau = \text{const} = \tau_0 \text{ (plane wire)}.$$

Then, the PDE system is reduced to an ordinary differential system of equations

$$(\omega - 2\tau_0)k = 0, \quad \left(-\tau^2 + \frac{k''}{k} + \frac{k^2}{2}\right)' = 0.$$

The first equation leads to $\omega = 2\tau_0$. If we impose the condition $\frac{k''}{k} > 0$, then the second differential equation becomes a second order Riccati equation

$$\frac{k''}{k} + \frac{k^2}{2} = C^2,$$

where C is a constant. The solitonic solution which satisfies the boundary condition is

$$k = 2C\,\text{sech}\,(C\xi).$$

It follows

$$q = 2C\,\text{sech}\,[C(s - 2\tau_0 t)]\exp(i\tau_0(s - s_0)),\ T(t) = -\tau_0^2 + C^2$$

and finnaly the solitonic solution of Schrödinger equation

$$Q = 2C\,\text{sech}\,[C(s - 2\tau_0 t)]\exp(i\tau_0(s - s_0) + (C^2 - \tau_0^2)t).$$

Going back to the last equations in Subsection 6.1, we find the Hasimoto surface $\vec{r} = \vec{r}(s,t)$ in parametric form (Theme!)

$$x(s,t) = \frac{s}{2} - \frac{C}{C^2 + \tau_0^2}\tanh\,(C\xi),$$

$$y(s,t) = -\frac{C}{C^2 + \tau_0^2}\,\text{sech}(C\xi)\,\cos[\tau_0 s + (C^2 - \tau_0^2)t],$$

$$z(s,t) = -\frac{C}{C^2 + \tau_0^2}\,\text{sech}(C\xi)\,\sin[\tau_0 s + (C^2 - \tau_0^2)t].$$

7.7 Evolution of implicit plane curves

Let $f : \mathbb{R}^2 \to \mathbb{R}$ be a smooth function. Let $C = f^{-1}(0)$ be an implicit plane curve, where 0 is a regular value of the function f, i.e., the normal vector field ∇f does not vanish on the curve. The curvature k of the curve C has the expression

$$k(x,y) = -\frac{f_{xx}f_y^2 - 2f_{xy}f_xf_y + f_{yy}f_x^2}{(f_x^2+f_y^2)^{\frac{3}{2}}}.$$

Let us suppose that $x = x(s), y = y(s), s \in I \subset R$ is a parametrization of the curve C, i.e., $f(x(s), y(s)) = 0, \forall s \in I$. Its evolution in time $C(t)$ (homotopy) is described by an equation

$$C(t) : F(x(s,t), y(s,t), t) = 0,$$

$$C(0) = C, \text{ i.e., } F(x,y,0) = f(x,y),\ (x,y) \in D \subset \mathbb{R}^2.$$

Denoting $\vec{r} = (x(s,t), y(s,t))$, by derivation with respect to the parameter t we obtain the relation

$$\langle \nabla F, \vec{r}_t \rangle + F_t = 0$$

or explicitly

$$\frac{\partial F}{\partial x}\frac{\partial x}{\partial t} + \frac{\partial F}{\partial y}\frac{\partial y}{\partial t} + \frac{\partial F}{\partial t} = 0.$$

If we accept the evolution along the vector field $\vec{X}$, i.e., $\frac{\partial}{\partial t}\vec{r} = \alpha\,\vec{X}$, then we find the PDE $F_t = -\alpha\langle \nabla F, \vec{X}\rangle$, with the unknown F, fixed by an initial condition.

Let $\vec{n} = \frac{\nabla F}{||\nabla F||}$ be the unit vector (versor) normal to the curve C. If we accept the evolution along the normal vector, $\frac{\partial}{\partial t}\vec{r} = \beta\,\vec{n}$, then we find the PDE $F_t = -\beta||\nabla F||$. For the fastest deformation it is taken $\beta = k$ and hence we arrive at the non-linear PDE

$$F_t = \frac{F_{xx}F_y^2 - 2F_{xy}F_xF_y + F_{yy}F_x^2}{F_x^2+F_y^2}$$

or

$$F_t = ||\nabla F||\,\mathrm{div}\left(\frac{\nabla F}{||\nabla F||}\right),$$

with the unknown $F(x,y,t)$.

Giving F, let us look at the relationship

$$\langle \nabla F, \vec{r}_t \rangle + F_t = 0$$

as an algebraic equation with the unknown $\vec{r}_t$. Finding a particular solution, the normal component, we obtain the normal velocity

$$\vec{r}_t = -F_t(x, y, t)\,\frac{\nabla F(x, y, t)}{\|\nabla F(x, y, t)\|^2}.$$

The general solution is

$$\vec{r}_t = -F_t(x, y, t)\,\frac{\nabla F(x, y, t)}{\|\nabla F(x, y, t)\|^2} + \lambda(x, y, t)X(x, y, t),$$

where $\langle \nabla F(x, y, t), X(x, y, t)\rangle = 0$, i.e., the vector field $X(x, y, t)$ is tangent to the curve $C(t)$, and $\lambda(x, y, t)$ is an arbitrary scalar field.

7.7.1 Deformations by infinitesimal transformation

Another problem is the deformation of a plane curve $C : f(x) = 0$, $x = (x^1, x^2)$ by an infinitesimal transformation.

Let us consider the differential system $\dot{x}(t) = X(x(t))$ on $U \subset \mathbb{R}^2$. A solution $\alpha = \alpha(t)$, $\alpha(0) = x$, $t \in (-\epsilon, \epsilon)$ generates a local diffeomorphism

$$T^t : U \subset \mathbb{R}^2 \to \mathbb{R}^2,\ T^t(x) = \Phi(t, x),\ t \in I.$$

This diffeomorphism is the solution of the operator differential equation

$$\frac{dT^t}{dt} = X \circ T^t,\ T^0 = id,$$

and is called *local flow on* $\mathbb{R}^2$ generated by the vector field X (*local group with one parameter of diffeomorphisms*). The linear approximation $x' = x + t\,X(x)$ of $T^t(x)$ is called *infinitesimal transformation generated by the vector field* X. This function of type $\varphi_t : \mathbb{R}^2 \to \mathbb{R}^2$ is indexed on the parameter t and is locally invertible. It follows $F = f \circ \varphi_t^{-1}$.

7.8 Self-evaluation problems

Problem 64. *Let $c(s) : I \to \mathbb{R}^2$ be a simple closed curve, of class C^2 related to curvilinear abscissa s, with the unit normal vector $\vec{n}(s)$. Let*

$\vec{v}(x^1, x^2) : \mathbb{R}^2 \to \mathbb{R}^2$ *be a continuous vector field. All these generate the functional (curvilinear integral of first type)*

$$I(c) = \int_{c(s)} \langle \vec{v}(c(s)), \vec{n}(s) \rangle ds.$$

Determine the first variation of the functional for a variation of the curve $c(s)$.

Problem 65. *There exists a curve* $\Gamma \subset \Omega = [0,1]^3$ *that joins the points* $(0,0,0)$ *and* $(1,1,1)$ *and that extremizes the functional*

$$I(\Gamma) = \int_\Gamma P(x,y,z)dx + Q(x,y,z)dy + R(x,y,z)dz\,?$$

Solution. We introduce the vector field $F = (P, Q, R)$. If F is an irrotational field, then the curvilinear integral is independent on the path (constant with respect to the path).

Let us find necessary conditions. Suppose that F is a rotational field. We fix the curve $\Gamma : x = x(t), y = y(t), z = z(t), t \in [a,b]$ of class C^1. Then the integral is written

$$I(\Gamma) = \int_a^b \langle F, \dot{\Gamma} \rangle \, dt.$$

Suppose that Γ is a curve which extremizes the functional and that $\Gamma_\varepsilon : x = x(t,\varepsilon), y = y(t,\varepsilon), z = z(t,\varepsilon), t \in [a,b], \varepsilon \in (-\delta, \delta)$ is its variation. We introduce the variation vector

$$\xi = \frac{\partial \Gamma_\varepsilon}{\partial \varepsilon}|_{\varepsilon=0} : u(t) = \frac{\partial x}{\partial \epsilon}(t;0), v(t) = \frac{\partial x}{\partial \epsilon}(t;0), w(t) = \frac{\partial x}{\partial \epsilon}(t;0),$$

which is cancelled at the ends. It follows the function (integral with one parameter)

$$I(\varepsilon) = \int_a^b \langle F(x(t,\varepsilon), y(t,\varepsilon), z(t,\varepsilon)), \dot{\Gamma}_\varepsilon \rangle dt.$$

The necessary condition is

$$0 = I'(0) = \int_a^b \Big(\langle \nabla_\xi F, \dot{\Gamma} \rangle + \langle F, \dot{\xi} \rangle \Big) \, dt.$$

Integrating by parts (last integral), we find

$$0 = I'(0) = \int_a^b \left(\langle \nabla_\xi F, \dot{\Gamma} \rangle - \langle \nabla_{\dot{\Gamma}} F, \xi \rangle \right) dt.$$

Since the vector ξ is arbitrary, it follows the necessary condition $\langle \nabla F, \dot{\Gamma} \rangle = \nabla_{\dot{\Gamma}} F$. It is not so easy to understand what this condition says.

Simpler The functional is transcribed

$$I(x(\cdot), y(\cdot), z(\cdot)) = \int_a^b (P(x,y,z)\dot{x} + Q(x,y,z)\dot{y} + R(x,y,z)\dot{z})\, dt.$$

The Lagrangian

$$L = P(x,y,z)\dot{x} + Q(x,y,z)\dot{y} + R(x,y,z)\dot{z}$$

has the derivatives

$$\frac{\partial L}{\partial x} = \frac{\partial P}{\partial x}\dot{x} + \frac{\partial Q}{\partial x}\dot{y} + \frac{\partial R}{\partial x}\dot{z}, \ \frac{\partial L}{\partial \dot{x}} = P$$

and similar for the other variables. It follows the Euler-Lagrange PDE

$$\left(\frac{\partial Q}{\partial x} - \frac{\partial P}{\partial y} \right) \dot{y} + \left(\frac{\partial R}{\partial x} - \frac{\partial P}{\partial z} \right) \dot{z} = 0$$

and the equations obtained by permutations. We find $\operatorname{rot} F \times \dot{\Gamma} = 0$, i.e., Γ must be a vortex line of the field F (field line of "rot F").

In short Denote $\bar{F} = P\,\bar{i} + Q\,\bar{j} + R\,\bar{k}$ and $\bar{T} = \dot{x}\,\bar{i} + \dot{y}\,\bar{j} + \dot{z}\,\bar{k}$. In case of the Lagrangian $L = \langle \bar{F}, \bar{T} \rangle$, the Euler-Lagrange system is

$$rot\, \bar{F} \times \bar{T} = \bar{0}.$$

Hence, either $rot\, \bar{F} = 0$ and the curvilinear integral is path independent or the curve is a vortex line.

Problem 66 (Elastic curves). *The elastic potential energy of a C^2 curve $c : [0, T] \to \mathbb{R}^2$ is the functional*

$$I(c(\cdot)) = \frac{1}{2} \int_0^T k^2(s) ds$$

(integral of the square of curvature). A curve of class C^2, which minimizes this functional, is called elastic curve. Determine all plane elastic curves of fixed length, which pass through two given points and having velocities prescribed at the ends.

Solution. Let $c(s) = (x(s), y(s))$ be a curve with unit velocity $\dot{c}$ and $\theta(s)$ the angle between the velocity $\dot{c}$ and the axis Ox. Then $k^2(s) = \dot{\theta}^2(s)$. Considering the restriction $\dot{x} = \cos\theta, \dot{y} = \sin\theta$, the problem of finding the elastic curves of length T can be solved optimizing the functional

$$J(c(\cdot)) = \int_0^T L(x, y, \theta, \dot{x}, \dot{y}, \dot{\theta})ds.$$

The Lagrangian

$$L = \frac{1}{2}\dot{\theta}^2 + \lambda_1(\dot{x} - \cos\theta) + \lambda_2(\dot{y} - \sin\theta)$$

generates the Euler-Lagrange ODEs

$$\ddot{\theta} = \lambda_1 \sin\theta - \lambda_2 \cos\theta,\ \dot{\lambda}_1 = 0, \dot{\lambda}_2 = 0.$$

Writing in a suitable form, we obtain the equation of the pendulum

$$\ddot{u} + a^2 \sin u = 0,$$

where

$$a^2 = \sqrt{\lambda_1^2 + \lambda_2^2},\ u = \theta + \pi - \alpha,\ \alpha = \operatorname{arctg}\frac{\lambda_2}{\lambda_1}.$$

The solutions can be written using elliptic functions.

Problem 67. *Look at the motion of the helix* $x = a\cos t, y = a\sin t, z = bt$, $t \in \mathbb{R}$ *along the direction of binormal vector field.*

Problem 68. *Study the motion of field lines of the vector field* (y, z, x) *along the direction of principal normal vector field.*

Problem 69. *Deform the astroid* $C : x^{\frac{2}{3}} + y^{\frac{2}{3}} = a^{\frac{2}{3}}$ *by the infinitesimal transformation associated to the vector field* $X = (xy, x^2 + y^2)$.

Problem 70. *Deform the strophoid* $C : x(x^2+y^2)+a(y^2-x^2) = 0$ *by the infinitesimal transformation associated to the normal vector field.*

Problem 71. *Deform the algebraic curves*

$$C_1 : F(x, y) = x^2(1 - x)(1 + x) - y^2 + 0.01 = 0,$$

$$C_2 : G(x, y) = x^2 - xy + y^4 + 0.0001 = 0$$

by the infinitesimal transformation associated to the vector field $X = (x + y, x - y)$.

Bibliography

[1] Q. Chang-Zheng, L. Yan-Yan, *Higher-dimensional integrable systems arising from motions of curves on $S^2(R)$ and $S^3(R)$*, Commun. Theor. Phys., Beijing, China, 50, 4 (2008), 841-843.

[2] R. L. Jerrard, D. Smets, *On the motion of a curve by its binormal curvature*, arXiv: 1109.5483v1[math.DG], 26 Sep. 2011.

[3] P. J. Olver, G. Sapiro, A. Tannenbaum, *Invariant geometric evolutions of surfaces and volumetric smoothing*, Manuscript, April 15, 1994.

[4] C. Rogers, W. K. Schief, *Backlund and Darboux Transformations, Geometry and Modern Applications in Soliton Theory*, Cambridge University Press, 2002.

[5] N. Sheridan, *Hamilton's Ricci Flow*, Honours Thesis, The University of Melbourne, Department of Mathematics and Statistics, November 2006.

[6] C. Udriste, *Geometric Dynamics*, Mathematics and Its Applications, 513, Kluwer Academic Publishers, Dordrecht, Boston, London, 2000.

[7] C. Udriste, L. Matei, *Lagrange-Hamilton Theories* (in Romanian), Monographs and Textbooks 8, Geometry Balkan Press, Bucharest, 2008.

[8] E. T. Whittaker, *A Treatise on The Analytical Dynamics of Particles & Rigid Bodies*, Cambridge University Press, 1989.

Chapter 8

MOTIONS OF SURFACES

Motto:

"Philosophy does not make a man wise
But the right mind, gained through through hard work,
Therefore do not look down on a ploughman:
Nobody knows what lies in a covered pot."

George Coşbuc - *The Philosophers and the Ploughmen*

Motion applies to various physical systems: objects, bodies, matter particles, radiation, radiation fields, radiation particles, curvature and space-time. One can also speak of motion of images, shapes and boundaries. So, the term motion, in general, signifies a continuous change in the position or configuration of a physical system in space.

In this Chapter, we analyse the motion of a parameterized surface, in space, introducing a third parameter, as an evolution parameter and specifying a PDE whose solution is the variation. In the context, we are considering a diffeomorphism and analyzing coordinate curves and coordinate surfaces by appropriate formulas. Another problem addressed is the flow of the steepest descent area. Such problems have direct applications in science and engineering, being related to evolutionary phenomena. Suggestions for further reading: [1]-[9].

8.1 Evolution of a surface in space

A large class of physical systems are shaped in terms of motion of surfaces (thin plates as free rigid bodies) in Euclidean space. Particularly, the theory in the following paragraphs includes the movement of thin plates through roto-translations.

Usually we use the free vector language combined with the position vector language. The evolution in time of a parameterized surface represented by the position vector $\vec{r} : D \subset \mathbb{R}^2 \to \mathbb{R}^3$, $\vec{r} = \vec{r}(u,v)$ is obtained by entering the time t as the third parameter indexing the value $\vec{r}(u,v)$ in the form $\vec{r}(u,v,t)$. Obviously this t can be a parameter in the vector equation of the surface or is introduced ad-hoc using an evolution PDE. In order to describe and understand evolution, we must first recall a few things related to the geometry of a surface.

We accept that by evolution of a surface in $\mathbb{R}^3$ we obtain a body (we avoid sliding along the surface, reparametrizations, extensions etc).

Our aim is to describe the kinematics of surfaces in $\mathbb{R}^3$ in terms of the differential geometry.

8.1.1 Geometry of a surface

We start with a surface described by a C^3 position vector $\vec{r} : D \subset \mathbb{R}^2 \to \mathbb{R}^3$, $\vec{r} = \vec{r}(u,v)$, with linear independent *partial velocities*, $\vec{r}_u, \vec{r}_v$, tangent to the surface and *normal versor field* $\vec{U} = \frac{\vec{r}_u \times \vec{r}_v}{||\vec{r}_u \times \vec{r}_v||}$. Often we notice $\mathcal{H}^2 = ||\vec{r}_u \times \vec{r}_v||^2 = EG - F^2$. It appears the moving frame $\{\vec{r}_u, \vec{r}_v, \vec{U}\}$, called *Gauss moving frame*, which satisfies

$$||\vec{U}|| = 1,\ \vec{U} \perp \vec{r}_u,\ \vec{U} \perp \vec{r}_v,\ \vec{U} \perp \vec{U}_u,\ \vec{U} \perp \vec{U}_v.$$

Note that the normal vector field $\vec{U}$ is a unit vector field but the partial velocities $\vec{r}_u, \vec{r}_v$ are not necessarily unit vectors.

Using the differential $d\vec{r} = \vec{r}_u du + \vec{r}_v dv$, we build the fundamental forms of the surface (quadratic forms):

(1) the *first fundamental form*,

$$I = \langle d\vec{r}, d\vec{r} \rangle_{\mathbb{R}^3} = E\,du^2 + 2F\,dudv + G\,dv^2,$$

with the coefficients

$$E = \langle \vec{r}_u, \vec{r}_u \rangle,\ F = \langle \vec{r}_u, \vec{r}_v \rangle,\ G = \langle \vec{r}_v, \vec{r}_v \rangle;$$

(2) the *second fundamental form*,

$$II = -\langle d\vec{r}, d\vec{U} \rangle_{\mathbb{R}^3} = \ell\,du^2 + 2m\,dudv + n\,dv^2$$

with the coefficients

$$\ell = -\langle \vec{r}_u, \vec{U}_u \rangle = \langle \vec{r}_{uu}, \vec{U} \rangle, \; n = -\langle \vec{r}_v, \vec{U}_v \rangle = \langle \vec{r}_{vv}, \vec{U} \rangle$$

$$m = -\langle \vec{r}_u, \vec{U}_v \rangle = -\langle \vec{r}_v, \vec{U}_u \rangle = \langle \vec{r}_{uv}, \vec{U} \rangle.$$

The fundamental forms determine the *Gauss curvature*

$$K = \frac{\ell m - n^2}{EG - F^2}$$

and the *mean curvature*

$$H = \frac{1}{2} \frac{G\ell - 2Fm + En}{EG - F^2}.$$

Although it is not easy to recognize here, we emphasize that the mean curvature is the half-trace of the second fundamental form with respect to the first fundamental form. The Gauss curvature is the quotient between determinant of the second fundamental form and the determinant of the first fundamental form.

Theorem 8.1.1 (Bonnet). *The fundamental forms determine the surface up to a roto-translation.*

With respect to the Gauss moving frame $\{\vec{r}_u, \vec{r}_v, \vec{U}\}$, a surface is characterized by *Gauss-Tzitzeica PDEs system*

$$\vec{r}_{uu} = \Gamma^1_{11}\, \vec{r}_u + \Gamma^2_{11}\, \vec{r}_v + \ell\, \vec{U}$$

$$\vec{r}_{uv} = \Gamma^1_{12}\, \vec{r}_u + \Gamma^2_{12}\, \vec{r}_v + m\, \vec{U}$$

$$\vec{r}_{vv} = \Gamma^1_{22}\, \vec{r}_u + \Gamma^2_{22}\, \vec{r}_v + n\, \vec{U}.$$

The coefficients Γ^i_{jk} determine the *Tzitzeica connection.* They do not have simple expressions because the Gaussian frame is not orthonormal. In the same frame, the *Weingarten PDEs* are

$$\vec{U}_u = \frac{mF - \ell G}{\mathcal{H}^2}\, \vec{r}_u + \frac{\ell F - mE}{\mathcal{H}^2}\, \vec{r}_v$$

$$\vec{U}_v = \frac{nF - mG}{\mathcal{H}^2}\, \vec{r}_u + \frac{mF - nE}{\mathcal{H}^2}\, \vec{r}_v.$$

Here we can judge in the following way: since $\vec{U}_v \perp \vec{U}$, i.e., the vector $\vec{U}_v$ is tangent to the surface, we can write

$$\vec{U}_v = a\vec{r}_u + b\vec{r}_v.$$

By scalar multiplication with $\vec{r}_u$ respectively $\vec{r}_v$, we obtain the algebraic system

$$aE + bF = -m,\ aF + bG = -n,$$

with the solution indicated above.

The complete integrability conditions of Gauss-Tzitzeica PDEs,

$$(\vec{r}_{uu})_v = (\vec{r}_{uv})_u,\ (\vec{r}_{uv})_v = (\vec{r}_{vv})_u$$

are known as *Mainardi-Codazzi PDEs*

$$\left(\frac{\ell}{H}\right)_v - \left(\frac{m}{H}\right)_u + \frac{\ell}{H}\,\Gamma^2_{22} - 2\,\frac{m}{H}\,\Gamma^2_{12} + \frac{n}{H}\,\Gamma^2_{11} = 0$$

$$\left(\frac{n}{H}\right)_u - \left(\frac{m}{H}\right)_v + \frac{\ell}{H}\,\Gamma^1_{22} - 2\,\frac{m}{H}\,\Gamma^1_{12} + \frac{n}{H}\,\Gamma^1_{11} = 0$$

or equivalent

$$\ell_v - m_u = \ell\Gamma^1_{12} + m(\Gamma^2_{12} - \Gamma^1_{11}) - n\Gamma^2_{11}$$

$$m_v - n_u = \ell\Gamma^1_{22} + m(\Gamma^2_{22} - \Gamma^1_{12}) - n\Gamma^2_{12}.$$

To formalize and simplify the writing, we need uniformization index notations

$$u^1 = u,\ u^2 = v,\ \vec{r}_u = \vec{r}_1, \vec{r}_v = \vec{r}_2,\ \vec{r}_{uu} = \vec{r}_{11},\ \vec{r}_{uv} = \vec{r}_{12},\ \vec{r}_{vv} = \vec{r}_{22},$$

$$I = g_{ij}du^i du^j,\ II = L_{ij}du^i du^j,\ i,j = 1,2.$$

We introduce the inverse (g^{ij}) of the metric matrix (g_{ij}) (matrix of the first fundamental form). Then $g_{ij}g^{jk} = \delta^k_i$.

The coefficients of the second fundamental form are the normal projections $L_{ij} = \langle \vec{r}_{ij}, \vec{U} \rangle$. In these notations, the Gauss curvature is $K = \det(g^{ij}L_{jk}) = \frac{\det(L_{ij})}{\det(g_{ij})}$, and the mean curvature is $H = \frac{1}{2}g^{ij}L_{ij}$, i.e., half-trace of L_{ij} with respect to the metric g_{ij}.

For bringing in the context of differential geometry, we define the Christoffel symbols (connection)

$$\Gamma^i_{jk} = \frac{1}{2}g^{il}\left(\frac{\partial g_{jl}}{\partial u^k} + \frac{\partial g_{kl}}{\partial u^j} - \frac{\partial g_{jk}}{\partial u^l}\right).$$

This connection is used to calculate covariant derivatives

$$\nabla_i A^j = \frac{\partial A^j}{\partial u^i} + \Gamma^j_{ki}A^k,\ \nabla_i a_j = \frac{\partial a_j}{\partial u^i} - \Gamma^k_{ji}a_k,$$

where A^j is a contravariant vector field, and respectively a_j is a covariant vector field. In fact, starting from now on, we use tensor ideas and notations; the indices rise with (g^{ij}) and lower with (g_{ij}).

The Gauss-Weingarten PDEs transcribe in matrix form

$$\frac{\partial}{\partial u^j}\begin{pmatrix} \vec{r}_i \\ \vec{U} \end{pmatrix} = \begin{pmatrix} \Gamma^k_{ij} & L_{ij} \\ -g^{ki}L_{ij} & 0 \end{pmatrix}\begin{pmatrix} \vec{r}_k \\ \vec{U} \end{pmatrix}.$$

8.1.2 Variation of a surface

To study the evolution of surfaces arbitrarily embedded in $\mathbb{R}^3$ (three dimensional space), a natural approach is to describe the surface purely intrinsically, that is, in terms of the surface's first fundamental form (metric) and second fundamental form.

Starting from a C^3 surface $\vec{r} = \vec{r}(u,v)$, $(u,v) \in D \subset \mathbb{R}^2$, we introduce a third parameter t (time), considering a variation

$$\vec{q} = \vec{q}(u,v,t),\ \vec{q}(u,v,0) = \vec{r}(u,v),\ (u,v) \in D \subset \mathbb{R}^2,\ t \in [0,\varepsilon),$$

of class C^3 and a given guided PDE of evolution. Obviously we can use the old notation $\vec{r}(u,v,t)$, but let's not forget what notation means in the new context (all functions depend on parameter t). For new notations, we introduce the decomposition

$$\frac{\partial}{\partial t}\vec{r} = V^i\ \vec{r}_i + W\ \vec{U},\ i = 1,2,$$

using Einstein summation convention.

Theorem 8.1.2. *The evolution PDEs of tangent vectors $\vec{r}_1(u,v,t)$, $\vec{r}_2(u,v,t)$ and of unit normal vector field $\vec{U}(u,v,t)$ are written somewhere in the matrix form*

$$\frac{\partial}{\partial t}\begin{pmatrix} \vec{r}_i \\ \vec{U} \end{pmatrix} = \begin{pmatrix} \nabla_i V^k - L^k_i W & \frac{\partial W}{\partial u^i} - V^j L_{ji} \\ -g^{ki}(\frac{\partial W}{\partial u^i} + V^j L_{ji}) & 0 \end{pmatrix}\begin{pmatrix} \vec{r}_k \\ \vec{U} \end{pmatrix}.$$

Proof. Using the decomposition

$$\frac{\partial}{\partial t}\vec{r} = V^i\ \vec{r}_i + W\ \vec{U},$$

we underline that

$$\frac{\partial}{\partial t}\vec{r}_i = \frac{\partial}{\partial t}\frac{\partial}{\partial u^i}\vec{r} = \frac{\partial}{\partial u^i}(V^j\ \vec{r}_j + W\ \vec{U})$$

$$= \frac{\partial V^j}{\partial u^i}\vec{r}_j + V^j\frac{\partial \vec{r}_j}{\partial u^i} + \frac{\partial W}{\partial u^i}\vec{U} + W\frac{\partial \vec{U}}{\partial u^i}$$

$$= ... = (\nabla_i V^j - L^j_i W)\vec{r}_j + \left(\frac{\partial W}{\partial u^i} - V^j L_{ji}\right)\vec{U}.$$

Let us now find the evolution equations for the normal vector field $\vec{U}$. The relation $\langle \vec{r}_i, \vec{U}\rangle = 0$ implies $\langle \frac{\partial \vec{r}_i}{\partial t}, \vec{U}\rangle = -\langle \vec{r}_i, \frac{\partial \vec{U}}{\partial t}\rangle$, and the relation $\langle \vec{U}, \vec{U}\rangle = 1$ implies $\langle \frac{\partial \vec{U}}{\partial t}, \vec{U}\rangle = 0$. Generically, $\frac{\partial \vec{U}}{\partial t} = \alpha^j\, \vec{r}_j + \beta\, \vec{U}$. The previous conditions impose $\beta = 0$, $\alpha^j = -g^{ji}\left(\frac{\partial \vec{U}}{\partial u^i} + V^k L_{ki}\right)$. □

Theorem 8.1.3. *The evolution PDEs of the metric g_{ij} are*

$$\frac{\partial g_{ij}}{\partial t} = \nabla_j V_i + \nabla_i V_j - 2L_{ij}W,\ V_i = g_{ij}V^j.$$

Proof. We first acknowledge that $g_{ij} = \langle \vec{r}_i, \vec{r}_j\rangle$ and hence

$$\frac{\partial g_{ij}}{\partial t} = \langle \frac{\partial \vec{r}_i}{\partial t}, \vec{r}_j\rangle + \langle \vec{r}_i, \frac{\partial \vec{r}_j}{\partial t}\rangle$$

$$= \langle \frac{\partial}{\partial u^i}\left(\frac{\partial \vec{r}}{\partial t}\right), \vec{r}_j\rangle + \langle \vec{r}_i, \frac{\partial}{\partial u^j}\left(\frac{\partial \vec{r}}{\partial t}\right)\rangle$$

$$= \langle \frac{\partial}{\partial u^i}\left(V^j\vec{r}_j + W\vec{U}\right), \vec{r}_j\rangle + ...$$

□

Theme Complete the previous calculations. Evaluate $\frac{\partial L_{ij}}{\partial t}$, $\frac{\partial g^{ij}}{\partial t}$, $\frac{\partial G}{\partial t}$, $\frac{\partial H}{\partial t}$.

Examples Study the evolutions

(1) $$\vec{r}(u, v, t) = e^{-t}(\cos u \cos v, \cos u \sin v, (1 - t)v);$$

(2) $$x = \operatorname{sech}(u + \frac{t}{2})\cos(u - v - \frac{t}{2}),$$

$$y = \operatorname{sech}(u + \frac{t}{2})\sin(u - v - \frac{t}{2}),$$

$$z = u + v - \frac{3t}{2}\tan(u + \frac{t}{2});$$

(3) $$\vec{r}(u, v, t) = e^{-t}(tu^2, uv, tv^2);$$

(4)
$$z^2 = x^2 + e^{-t}y^2;\; z = \frac{1}{4}x^2 - ty^2; z = xy\cosh\, t;\; z = (\ln\cos x - \ln\cos y)\sinh\, t$$

through fundamental forms.

8.2 Motion of a surface along normal vector

Studies of the motion of surfaces and interfaces occur in several scientific disciplines, including differential geometry, water waves, crystal growth, combustion, and gas dynamics.

We start with a surface described by the position vector $\vec{r} : D \subset \mathbb{R}^2 \to \mathbb{R}^3$, $\vec{r} = \vec{r}(u,v)$, of class C^3, with linear independent partial velocities $\vec{r}_u, \vec{r}_v$ tangent to the surface and the unit normal vector field $\vec{U} = \frac{\vec{r}_u \times \vec{r}_v}{||\vec{r}_u \times \vec{r}_v||}$. All together determine the Gauss moving frame $\{\vec{r}_u, \vec{r}_v, \vec{U}\}$, and the mean curvature

$$H = \frac{1}{2}\frac{G\ell - 2Fm + En}{EG - F^2}.$$

A time evolution of the surface is described by a differentiable variation

$$\vec{q} = \vec{q}(u,v,t),\; \vec{q}(u,v,0) = \vec{r}(u,v),\; (u,v) \in D \subset \mathbb{R}^2,\; t \in [0,\varepsilon),$$

determined as solution of a given evolution PDEs system.

Let us accept that that the evolution is described by the Cauchy problem

$$\partial_t \vec{q}(u,v,t) = \vec{U}(u,v),\; \vec{q}(u,v,0) = \vec{r}(u,v)$$

(motion along normal vector field of the initial surface). Then the solution is

$$\vec{q}(u,v,t) = \vec{r}(u,v) + t\,\vec{U}(u,v),\; (u,v) \in D,\; t \in \mathbb{R}.$$

Remark 8.2.1. *Particularly, we accept that we start with the graph of equation $z = z(x,y)$, attached to a function $z(x,y)$ of class C^2. Then $\vec{r}(x,y) = (x,y,z(x,y))$ and hence*

$$\vec{r}_x = (1,0,z_x),\; \vec{r}_y = (1,0,z_y),\; \vec{U} = \frac{-z_x\vec{\imath} - z_y\vec{\jmath} + \vec{k}}{\sqrt{1 + z_x^2 + z_y^2}},$$

$$K = \frac{z_{xx}z_{yy} - z_{xy}^2}{(1 + z_x^2 + z_y^2)^2}, \; 2H = \frac{\left(1 + z_y^2\right) z_{xx} - 2z_x z_y z_{xy} + \left(1 + z_x^2\right) z_{yy}}{(1 + z_x^2 + z_y^2)^{3/2}}.$$

The time evolution of the surface, along the normal vector, is described by a differentiable variation

$$w = w(x, y, t), \; w(x, y, 0) = z(x, y),$$

represented by the position vector

$$\vec{q}(x, y, t) = (x, y, w(x, y, t))$$

that must be a solution of an imposed evolution PDE.

The most known evolution along the normal is described by the Cauchy problem (PDEs + initial conditions)

$$(PDE) \qquad \partial_t \vec{q}(u, v, t) = -H(u, v, t)\, \vec{U}(u, v, t), \; \vec{q}(u, v, 0) = \vec{r}(u, v),$$

where H is the mean curvature (normal scaled by opposite of mean curvature).

Surfaces that minimize area subject to various constraints have long been studied. Much of the inspiration for these studies has come from physical systems involving surface tension: soap films, soap bubbles, capillarity, biological cell structure, and others. So far, mathematical investigations have been mostly confined to the equilibrium states of the mentioned systems, with some study of the evolution of non-parametric surfaces.

Let us show that the evolution along the vector $-H\,\vec{U}$ is actually a flow that decreases the area.

Theorem 8.2.1. *Let be the area functional (transcribed as double integral with one parameter)* $\omega(t) = \int_{[0,1]^2} ||\vec{q}_u \times \vec{q}_v||(u, v, t)\, dudv$, *where* $\vec{q}(u, v, t)$ *is an arbitrary variation of the surface* $\vec{q}(u, v, 0) = \vec{r}(u, v)$, *which satisfies boundary problem*

$$\partial_t \vec{q}(u, v, t) = \vec{V}(u, v, t), \; \vec{V}(u, v, t)|_{(u,v) \in \partial[0,1]^2} = 0,$$

with $\vec{V}$ *normalized by the condition* $\int_{[0,1]^2} ||\vec{V}(u, v, t)||^2\, dudv = 1$. *Then the decreasing rate* $-\frac{d\omega}{dt}$ *of the area is maximal if* $\vec{V}$ *is collinear to* $\vec{U}$.

Proof. Firstly, we compute the partial derivative

$$\partial_t ||\vec{q}_u \times \vec{q}_v|| = \partial_t \langle \vec{q}_u \times \vec{q}_v, \vec{q}_u \times \vec{q}_v \rangle^{\frac{1}{2}}$$

$$= \frac{1}{||\vec{q}_u \times \vec{q}_v||} \langle \partial_t \vec{q}_u \times \vec{q}_v + \vec{q}_u \times \partial_t \vec{q}_v, \vec{q}_u \times \vec{q}_v \rangle$$

$$= \langle \vec{q}_{tu} \times \vec{q}_v + \vec{q}_u \times \vec{q}_{tv},\ \vec{U} \rangle = \langle \vec{V}_u \times \vec{q}_v + \vec{q}_u \times \vec{V}_v,\ \vec{U} \rangle.$$

It follows

$$\frac{d\omega}{dt} = \int_{[0,1]^2} \partial_t ||\vec{q}_u \times \vec{q}_v||\ dudv$$

$$= \int_{[0,1]^2} \langle \vec{V}_u \times \vec{q}_v + \vec{q}_u \times \vec{V}_v,\ \vec{U} \rangle\ dudv$$

$$= \int_{[0,1]^2} \langle \partial_u (\vec{V} \times \vec{q}_v) - \vec{V} \times \vec{q}_{uv} + \partial_v (\vec{V} \times \vec{q}_u) - \vec{V} \times \vec{q}_{uv},\ \vec{U} \rangle\, dudv.$$

The integrand can be written in the form

$$\partial_u \langle \vec{V} \times \vec{q}_v,\ \vec{U} \rangle - \langle \vec{V} \times \vec{q}_v,\ \vec{U}_u \rangle + \partial_v \langle \vec{V} \times \vec{q}_u,\ \vec{U} \rangle - \langle \vec{V} \times \vec{q}_u,\ \vec{U}_v \rangle - 2\langle \vec{V} \times \vec{q}_{uv},\ \vec{U} \rangle.$$

That's why we find

$$\frac{d\omega}{dt} = \int_{[0,1]^2} \Big(2\langle \vec{V} \times \vec{q}_{uv},\ -\vec{U} \rangle - \langle \vec{V} \times \vec{q}_v,\ \vec{U}_u \rangle - \langle \vec{V} \times \vec{q}_u,\ \vec{U}_v \rangle \Big)\, dudv$$

$$= \int_{[0,1]^2} \langle 2\vec{U} \times \vec{q}_{uv} + \vec{U}_u \times \vec{q}_v + \vec{U}_v \times \vec{q}_u,\ \vec{V} \rangle\ dudv$$

$$= \int_{[0,1]^2} \langle 2\vec{U} \times \vec{q}_{uv} - 2H ||\vec{q}_u \times \vec{q}_v||\, \vec{U},\ \vec{V} \rangle\ dudv$$

$$= \int_{[0,1]^2} \langle -2H ||\vec{q}_u \times \vec{q}_v||\, \vec{U},\ \vec{V} \rangle\ dudv + \int_{[0,1]^2} \langle 2\vec{U} \times \vec{q}_{uv},\ \vec{V} \rangle\ dudv$$

$$\leq \int_{[0,1]^2} \langle -2H ||\vec{q}_u \times \vec{q}_v||\, \vec{U},\ \vec{V} \rangle\ dudv,$$

with equality only if $\vec{U}$ is collinear to $\vec{V}$.

Vranceanu simplified variant We start with the variation along the initial normal

$$\vec{q}(u,v,t) = \vec{r}(u,v) + t\,\varphi(u,v)\vec{U}(u,v),\ (u,v) \in D,\ t \in (-\delta, \delta).$$

Since

$$\vec{q}_u(u,v,t) = \vec{r}_u(u,v) + t\,(\varphi_u(u,v)\vec{U}(u,v) + \varphi(u,v)\vec{U}_u(u,v))$$

$$\vec{q}_v(u,v,t) = \vec{r}_v(u,v) + t\,(\varphi_v(u,v)\vec{U}(u,v) + \varphi(u,v)\vec{U}_v(u,v)),$$

for t enough small, the normal variation is also a surface (check the regularity condition). Denoting g_{ij} and h_{ij}, the first fundamental form and the second fundamental form of the surface $\vec{r} = \vec{r}(u,v)$, we find the first fundamental form of the variation surface

$$G_{ij} = g_{ij} - 2\,t\,\varphi\,h_{ij} + \mathcal{O}_{ij}(t^2),\ i,j = 1,2.$$

This remains positive definite for t enough small, independent of φ. The area of a compact subset of the variation (transcribed like a double integral with a parameter) is

$$\omega(t) = \int_{[0,1]^2} \sqrt{\det(G_{ij})}\,(u,v,t)\,dudv.$$

It follows

$$0 = \partial_t\omega(t)|_{t=0} = \int_{[0,1]^2} \left(\frac{1}{2\sqrt{\det(G_{ij})}}\,\partial_t \det(G_{ij})\right)_{t=0} dudv$$

$$= -\int_{[0,1]^2} \varphi\,\frac{1}{\det(g_{ij})}(h_{11}g_{22} + h_{22}g_{11} - 2h_{12}g_{12})\sqrt{\det(g_{ij})}\,dudv$$

$$= -\int_{[0,1]^2} 2\,\varphi\,H\sqrt{\det(g_{ij})}\,dudv.$$

If we select $\varphi = H$ (mean curvature) and the inward orientation of $\vec{U}$, then we obtain the result: *a condition required that the area of a compact subset of the surface $\vec{r} = \vec{r}(u,v)$ be smaller than or equal to the areas of normal variations is the non-cancellation of the mean curvature.* □

The surfaces with $H = 0$ are called *minimal surfaces* (ex., soap balloons, catenoid, helicoid, Enneper surface, Costa's minimal surface, and the Gyroid).

Example 8.2.2 (Shrinking sphere). Let's consider the sphere

$$\vec{r}(u,v) = (R_0 \sin u \cos v, R_0 \sin u \sin v, R_0 \cos u)$$

submersed in the body (spherical crown)

$$\vec{q}(u,v,t) = R(t)(\sin u \cos v, \sin u \sin v, \cos u).$$

In this case we have $\vec{N} = -\frac{1}{R}\,\vec{q}$, $H = \frac{1}{R}$ and the evolution equation is reduced to $\frac{dR}{dt} = -\frac{1}{R}$. It follows $R = (R_0^2 - 2t)^{\frac{1}{2}}$ and

$$\vec{q}(u,v,t) = (R_0^2 - 2t)^{\frac{1}{2}}(\sin u \cos v, \sin u \sin v, \cos u).$$

This new sphere tends to origin when $t \to \frac{1}{2}R_0^2$.

8.3 Minimal surface PDE

Let $f : \mathbb{R}^2 \to \mathbb{R}$ be a C^2 function. We consider the compact surface

$$\Sigma = \{(x,y,f(x,y))|\ (x,y) \in \Omega\} \subset \mathbb{R}^3,$$

whose area is

$$A(\Sigma) = I(f(\cdot)) = \int_\Omega \sqrt{1 + ||\nabla f||^2}\, dxdy.$$

The Euler-Lagrange PDE associated to this functional (the PDE of minimal surface, i.e., $H = 0$) is (Theme!)

$$\operatorname{div}\left(\frac{\nabla f}{\sqrt{1 + ||\nabla f||^2}}\right) = 0$$

or, explicitly,

$$\left(1 + f_y^2\right) f_{xx} - 2 f_x f_y f_{xy} + \left(1 + f_x^2\right) f_{yy} = 0,$$

with the unknown function $f(x,y)$. Examples include the catenoid, helicoid, Enneper surface, Costa's minimal surface, and the Gyroid.

This is one of the most studied nonlinear equations with partial derivatives. The type of this equation is elliptical, because its linearization is elliptical. Indeed, if a differentiable variation $f(x,y,\varepsilon)$ satisfies this equation and the Cauchy condition $f(x,y,0) = f(x,y)$, partially deriving with respect to ε and denoting $\partial_\varepsilon f|_{\varepsilon=0} = \xi$, we find the linearization (equation in variations)

$$2 f_y f_{xx} \xi_y + (1 + f_y^2)\xi_{xx} - 2 f_y f_{xy} \xi_x - 2 f_x f_{xy} \xi_y - 2 f_x f_y \xi_{xy}$$

$$+2f_x f_{yy}\xi_x + (1+f_x^2)\xi_{yy} = 0,$$

(the unknown is ξ) which is found to be elliptical.

The associated parabolic PDE

$$\partial_t f = \sqrt{1+||\nabla f||^2}\ \mathrm{div}\left(\frac{\nabla f}{\sqrt{1+||\nabla f||^2}}\right),\ f = f(x,y,t)$$

is called *mean curvature flow* or *negative gradient flow* of the area functional. In the limits $|f_x| \ll 1$, $|f_y| \ll 1$, i.e., the surface is nearly planar with its normal nearly parallel to the Oz axis, this reduces to a diffusion PDE $f_t = \nabla^2 f$.

Because the problem of linearizing a nonlinear PDE is not easy to guess, we repeat the above construction for another equation.

Problem 72. *Linearize the nonlinear PDE*

$$\frac{\partial u}{\partial t}(x,t) = u^2(x,t)\,\frac{\partial^2 u}{\partial x^2}(x,t),$$

around a solution $u(x,t)$.

Solution. Accepting that a differentiable variation $u(x,t,\varepsilon)$, of $u(x,t,0) = u(x,t)$, satisfies this PDE, partially deriving with respect to ε and denoting $\partial_\varepsilon u|_{\varepsilon=0} = \xi$, we find the linearization (equation in variations)

$$\frac{\partial \xi}{\partial t}(x,t) = u^2(x,t)\,\frac{\partial^2 \xi}{\partial x^2}(x,t) + 2u(x,t)\,\frac{\partial^2 u}{\partial x^2}(x,t)\,\xi(x,t),$$

(the unknown is ξ) around a solution $u(x,t)$ of the initial PDE.

8.4 Evolution of implicit surfaces

Let $F : \mathbb{R}^3 \to \mathbb{R}$ be a smooth function. Suppose that 0 is a regular value of the function F, i.e., the normal vector field ∇F does not vanish on the set $S = F^{-1}(0)$. Then $S = F^{-1}(0)$ is an implicit and regular surface (given as the set of zeroes of F). The implicit surface S is a constant level set of the function F. Generally, the constant level sets of the function F are given by $M_c : F(x,y,z) = c$.

The Gauss curvature of the surface S has the expression

$$K = [[F_3(F_3F_{11} - 2F_1F_{13}) + F_1^2F_{33}][F_3(F_3F_{22} - 2F_2F_{23}) + F_2^2F_{33}$$

$$- (F_3(-F_1F_{23} + F_3F_{12} - F_2F_{13}) + F_1F_2F_{33})^2\big] \left[F_3^2(F_1^2+F_2^2+F_3^2)^2\right]^{-1},$$

where the indices mean partial derivatives (1 for the derivative with respect to x, 2 for the derivative with respect to y etc).

The mean curvature of an implicit surface can be calculated by using the gradient

$$\nabla F = \left(\frac{\partial F}{\partial x}, \frac{\partial F}{\partial y}, \frac{\partial F}{\partial z}\right)$$

and the Hessian matrix

$$\text{Hess}(F) = \begin{pmatrix} F_{11} & F_{12} & F_{13} \\ F_{21} & F_{22} & F_{23} \\ F_{31} & F_{32} & F_{33} \end{pmatrix}.$$

Indeed, the mean curvature is

$$H = \frac{\nabla F \ \text{Hess}(F) \ \nabla F^T - \|\nabla F\|^2 \ Tr(\text{Hess}(F))}{2\|\nabla F\|^3}.$$

A more simplified form is as the divergence of the unit normal vector field, i.e.,

$$H = -\frac{1}{2} \ \nabla \cdot \frac{\nabla F}{\|\nabla F\|}.$$

Let us suppose that $x = x(u,v), y = y(u,v), z = z(u,v), (u,v) \in D \subset \mathbb{R}^2$ is a parametrization of the surface S, i.e.,

$$F(x(u,v), y(u,v), z(u,v)) = 0, \ \forall (u,v) \in D.$$

Its differentiable evolution in time $S(t)$ is described by an implicit equation

$$S(t) : G(x(u,v,t), y(u,v,t), z(u,v,t), t) = 0, \ (u,v) \in D, \ t \in J, \ S(0) = S,$$

i.e.,

$$G(x,y,z,0) = F(x,y,z), \ (x,y,z) \in \Omega \subset \mathbb{R}^3.$$

Denoting $\vec{r} = (x(u,v,t), y(u,v,t), z(u,v,t))$, by derivation with respect to t, we find that the function G verifies the relation

$$\langle \nabla G, \vec{r}_t \rangle + G_t = 0$$

or explicitly

$$\frac{\partial G}{\partial x}\frac{\partial x}{\partial t} + \frac{\partial G}{\partial y}\frac{\partial y}{\partial t} + \frac{\partial G}{\partial z}\frac{\partial z}{\partial t} + \frac{\partial G}{\partial t} = 0.$$

If we impose the evolution along the direction of the vector field $\vec{X}$, i.e., $\frac{\partial}{\partial t}\vec{r} = \alpha\,\vec{X}$, then we find the PDE $G_t = -\alpha(\nabla G, \vec{X})$, with the unknown G, fixed by an initial condition.

Let $\vec{n} = \frac{\nabla G}{||\nabla G||}$ be the versor field normal to the surface $S(t)$. If we accept the evolution along the normal vector, i.e., $\frac{\partial}{\partial t}\vec{r} = \beta\,\vec{n}$, then we find the PDE $G_t = -\beta||\nabla G||$. For deeper geometric properties, some authors take $\beta = K$ (curvature), while others take $\beta = H$ (mean curvature), reaching interesting second order nonlinear PDEs with the unknown function G.

8.4.1 Normal velocity

A time evolving implicit surface is defined by a differentiable function F whose parameters change over time. More formally the surface is

$$\sigma(t) = \{x = (x^1, x^2, x^3) \in \mathbb{R}^3 \mid F(x,t) = 0\}, \ \text{for } t \in J.$$

By computing the total time derivative of the composed function, we obtain information on the velocity of the points on the surface,

$$\frac{d}{dt}F(x,t) = \langle \nabla F(x,t), \dot{x}(t)\rangle + \frac{\partial F}{\partial t}(x,t).$$

For points on the surface $\sigma(t)$ this derivative has to be zero. Therefore

$$\langle \nabla F(x,t), \dot{x}(t)\rangle = -\frac{\partial F}{\partial t}(x,t).$$

Remark 8.4.1. *The above explanations show a way to animate the implicit surface $\sigma(t)$. We define a velocity field, the only unknown quantity in the previous equation, and integrate it over time.*

Giving $F(x,t)$, the previous equation in the unknown $\dot{x}(t)$ has an infinity of solutions. One of these solutions is that of minimum norm (projection of $\dot{x}(t)$ on $\nabla F(x,t)$), i.e.,

$$\dot{x}(t) = -\frac{\partial F}{\partial t}(x,t)\ \frac{\nabla F(x,t)}{||\nabla F(x,t)||^2} = v_n(x,t)\ n(x,t),$$

where $n(x,t) = \frac{\nabla F(x,t)}{||\nabla F(x,t)||}$. This equation provides only partial information on the velocity, namely, only its variation in the direction of the gradient ∇F. The tangential component can be arbitrary since motion within the surface does not change the shape. This reflects the fact that an implicit surface does not have a unique parametrization.

The general solution is

$$\dot{x}(t) = -\frac{\partial F}{\partial t}(x,t)\ \frac{\nabla F(x,t)}{\|\nabla F(x,t)\|^2} + \lambda(x,t)X(x,t),$$

where $\langle \nabla F(x,t), X(x,t)\rangle = 0$, i.e., the vector field $X(x,t)$ is tangent to the surface $\sigma(t)$, and $\lambda(x,t)$ is an arbitrary scalar field.

8.4.2 Tangential velocity

What we propose in this subsection is to require that the normal vector field does not vary over time. For translational motions this is clearly the case. Again we emphasize that in Subsection 4.1, the velocity vector field $\dot{x}(t)$ has an arbitrary tangent component. Our goal is to chose one that is close to the hypothetical motion of a surface particle and which is exact for translational motions. Mathematically our condition states that the time derivative of the normalized gradient should vanish on the surface, i.e.,

$$\frac{d}{dt}\frac{\nabla F}{\|\nabla F\|} = 0.$$

To compute this derivative we use the fact that a derivative "D" of the normalized vector $n = \nabla F/\|\nabla F\|$ is the projection of the derivative onto the plane normal to the vector, i.e.,

$$Dn = \frac{1}{\|\nabla F\|}\ P_n\ D\,\nabla F,$$

where $P_n = I_3 - n \otimes n$ is the projection operator to the plane normal to the vector n and I_3 is the 3×3 identity matrix. Similar to the case of the normal velocity, we have that the (total) time derivative is equal to sum of two terms

$$\frac{d}{dt}\frac{\nabla F}{\|\nabla F\|} = \frac{\partial}{\partial t}\frac{\nabla F}{\|\nabla F\|} + \langle \nabla\ \frac{\nabla F}{\|\nabla F\|}, \dot{x}\rangle.$$

On the other hand,

$$\frac{\partial}{\partial t}\frac{\nabla F}{\|\nabla F\|} = \frac{1}{\|\nabla F\|}\ P_n\ \frac{\partial}{\partial t}\nabla F,\ \ \langle \nabla \frac{\nabla F}{\|\nabla F\|}, \dot{x}\rangle = \frac{1}{\|\nabla F\|}\ P_n\ \nabla\nabla^T F.$$

Hence

$$0 = \frac{d}{dt}\frac{\nabla F}{\|\nabla F\|} = \frac{1}{\|\nabla F\|} P_n \left(\frac{\partial}{\partial t} \nabla F + \nabla \nabla^T F \, \dot{x} \right).$$

Solving, we obtain the tangential component of the velocity

$$P_n \, H_F \, \dot{x} = -P_n \, \frac{\partial}{\partial t} \nabla F,$$

where $H_F = \nabla \nabla^T F$ is the Hessian of the function F.

8.4.3 Deformation of surfaces by an infinitesimal transformation

An important problem is the deformation of the surface $S : F(x) = 0$, $x = (x^1, x^2, x^3) \in \mathbb{R}^3$ by an infinitesimal transformation associated to a flow.

We consider the differential system $\dot{x}(t) = X(x(t))$ on $U \subset \mathbb{R}^3$. A solution $\alpha = \alpha(t)$, $\alpha(0) = x$, $t \in (-\epsilon, \epsilon)$ generates a local diffeomorphism

$$T^t : U \subset \mathbb{R}^3 \to \mathbb{R}^3, \, T^t(x) = \Phi(t, x), \, t \in I.$$

This diffeomorphism is the solution of operatorial differential equation

$$\frac{dT^t}{dt} = X \circ T^t, \, T^0 = id,$$

and is called *local flow on* $U \subset \mathbb{R}^3$ generated by the vector field X (*local group with one parameter of diffeomorphisms*). The linear approximation $x' = x + t\, X(x)$ of $T^t(x)$ is called *infinitesimal transformation generated by* X. Locally this is an invertible function of type $\varphi_t : \mathbb{R}^3 \to \mathbb{R}^3$ indexed by t. It follows the deformed function $G = F \circ \varphi_t^{-1}$.

Remark 8.4.2. *(1) Similarly we can talk about the movement of a body (three dimensional Riemannian manifold)* $\vec{r} = \vec{r}(u, v, w)$ *in the time* t*, represented by a differentiable variation* $\vec{q} = \vec{q}(u, v, w, t)$ *on* $\mathbb{R}^4$ *(see also coordinate transformations indexed by the parameter* t*). The general theory includes the movement of free rigid bodies through roto-translations. A free rigid body has six degrees of freedom (3 translations + 3 rotations).*

(2) Generally, for a given Riemannian manifold $(M, g(x))$*, we can introduce the evolution* $g(x, t)$ *of the Riemannian metric through interesting evolution PDEs (as for example Ricci flow).*

8.5 Self-evaluation problems

Problem 73. *Let $D \subset \mathbb{R}^2$ be a compact set and a C^2 surface S represented by the position vector $\vec{r} = \vec{r}(u,v) : D \to \mathbb{R}^3$. Denote by $\vec{n}$ the unit normal vector of the surface. Let $\vec{v}(x^1, x^2, x^3) : \mathbb{R}^3 \to \mathbb{R}^3$ be a continuous vector field. All these generate the functional (surface integral of the first type)*

$$I(S) = \iint_S \langle \vec{v}(\vec{r}(u,v)), \vec{n}(u,v) \rangle d\sigma.$$

Determine the first variation of the functional at a variation of the surface S.

Problem 74. *There exists a surface $\Sigma \subset \mathbb{R}^3$ whose boundary $\partial\Omega$ is closed, fixed, and which extremizes the functional (surface integral of the second type)*

$$I(\Sigma) = \int_\Sigma P(x,y,z)dydz + Q(x,y,z)dzdx + R(x,y,z)dxdy\,?$$

Solution. Suppose $\Sigma : x = x(u,v), y = y(u,v), z = z(u,v), (u,v) \in D = [a,b] \times [c,d]$, of class C^2. Then the surface integral functional is reduced to

$$\int_D \begin{vmatrix} P & Q & R \\ x_u & y_u & z_u \\ x_v & y_v & z_v \end{vmatrix} dudv.$$

We use the Lagrangian

$$L = \begin{vmatrix} P & Q & R \\ x_u & y_u & z_u \\ x_v & y_v & z_v \end{vmatrix}.$$

The Euler-Lagrange PDEs are of the form

$$\frac{\partial L}{\partial x} - D_u \frac{\partial L}{\partial x_u} - D_v \frac{\partial L}{\partial x_v} = 0;$$

similarly, for y and z. On the other hand

$$\frac{\partial L}{\partial x} = \begin{vmatrix} \frac{\partial P}{\partial x} & \frac{\partial Q}{\partial x} & \frac{\partial R}{\partial x} \\ x_u & y_u & z_u \\ x_v & y_v & z_v \end{vmatrix}, \quad \frac{\partial L}{\partial x_u} = \begin{vmatrix} P & Q & R \\ 1 & 0 & 0 \\ x_v & y_v & z_v \end{vmatrix}, \quad \frac{\partial L}{\partial x_v} = \begin{vmatrix} P & Q & R \\ x_u & y_u & z_u \\ 1 & 0 & 0 \end{vmatrix}.$$

It follows the first Euler-Lagrange PDE

$$\begin{vmatrix} \frac{\partial P}{\partial x} & \frac{\partial Q}{\partial x} & \frac{\partial R}{\partial x} \\ x_u & y_u & z_u \\ x_v & y_v & z_v \end{vmatrix} - D_u \begin{vmatrix} P & Q & R \\ 1 & 0 & 0 \\ x_v & y_v & z_v \end{vmatrix} - D_v \begin{vmatrix} P & Q & R \\ x_u & y_u & z_u \\ 1 & 0 & 0 \end{vmatrix} = 0.$$

To find a conclusion using these types of PDEs is a difficult task.

To simplify, we are looking for the surface Σ as $z = z(x, y)$. Then we have a unique Euler-Lagrange PDE

$$P_x + Q_y + R_z = 0,$$

i.e., the vector field (P, Q, R) must be solenoidal (there exists a vector potential A such that $(P, Q, R) = \nabla \times A$).

In short Denote $\bar{F} = (P, Q, R)$. Let us consider the Lagrangian $L = (\bar{F}, \bar{r}_u, \bar{r}_v)$, mixt product. The Euler-Lagrange system is

$$div\bar{F}\,(\bar{r}_u \times \bar{r}_v) = \bar{0}$$

(a scalar multiplied by a vector). Hence, either $div\bar{F} = 0$ and the integral surface functional is constant (modulo the sign), or the Euler-Lagrange system has no solution.

Problem 75 (Generalization). *Let* $\omega = a_{i_1 \ldots i_k}(x)\, dx^{i_1} \wedge \ldots \wedge dx^{i_k}$, $1 \leq i_1 < \ldots < i_k \leq n$, *be a k-form of the class* C^2 *in* $\mathbb{R}^n$. *We look for the extremals, relative to k-dimensional manifolds* Ω, *for the integral functional*

$$I(\Omega) = \int_\Omega \omega = \int_\Omega a_{i_1 \ldots i_k}(x)\, dx^{i_1} \wedge \ldots \wedge dx^{i_k}.$$

Solution. Using for Ω a parametrization $x = x(u)$, $u = (u_1, \ldots, u_k) \in D \subset \mathbb{R}^n$, we obtain the Lagrangian

$$L\left(x(u), \frac{\partial x^1}{\partial u^1}, \ldots \frac{\partial x^n}{\partial u^k}\right) = a_{i_1 \ldots i_k}(x(u))\, \frac{\partial(x^{i_1}, \ldots, x^{i_k})}{\partial(u^1, \ldots u^k)},$$

$1 \leq i_1 < \ldots < i_k \leq n$.

In order to write the corresponding Euler - Lagrange equations, one needs the following detail: a p-form $\lambda = c_{i_1 \ldots i_p}(x)\, dx^{i_1} \wedge \ldots \wedge dx^{i_p}$, $1 \leq i_1 < \ldots < i_p \leq n$, is equivalent to a p-skew-symmetric tensor field $T(\lambda) = c_{i_1 \ldots i_p}$.

Let us return to writing Euler - Lagrange equations

$$\frac{\partial L}{\partial x^i} - \frac{\partial}{\partial u^j}\left(\frac{\partial L}{\partial x^i_{u^j}}\right) = 0, \; j = 1, ..., k, \; i = 1, ..., n.$$

Consider the exterior differential as a $(k+1)$-form

$$d\omega = \frac{\partial a_{i_1 \dots i_k}}{\partial x^j}\, dx^{i_1} \wedge \dots \wedge dx^{i_k} \wedge dx^j = b_{i_1 \dots i_k i_{k+1}}\, dx^{i_1} \wedge \dots \wedge dx^{i_k} \wedge dx^{k+1},$$

the 1-forms $\alpha^j = \sum_{i=1}^{n} x^i_{u^j}\, dx^i$ and the k-form

$$\alpha = \alpha^1 \wedge \dots \wedge \alpha^k = \sigma_{j_1 \dots j_k}\, dx^{j_1} \wedge \dots \wedge dx^{j_k}.$$

Using the tensors $T(d\omega) = b_{i_1 \dots i_k i_{k+1}}$ and $T(\alpha) = \sigma_{j_1 \dots j_k}$, the Euler - Lagrange equations can be written

$$\sum_{i_1 \dots i_k} b_{i, i_1, \dots, i_k} \sigma_{i_1, \dots, i_k} = 0.$$

Problem 76. *Analyse the motion of Tzitzeica surface $S : xyz = 1$ along the vector field (y, z, x).*

Solution. Firstly, we must find the general solution of the PDE

$$F_t + yF_x + zF_y + xF_z = 0.$$

We use the equation $\varepsilon^3 = 1$, whose solutions are $\varepsilon_0, \varepsilon_1, \varepsilon_2$. Since the characteristic system

$$\frac{dt}{1} = \frac{dx}{y} = \frac{dy}{z} = \frac{dz}{x}$$

admits the first integrals (*the passage into the complex numbers is understood*)

$$e^{-t}(x+y+z) = c_1, \; e^{-\varepsilon_2 t}(x+\varepsilon_1 y+\varepsilon_1^2 z) = c_2, \; e^{-\varepsilon_1 t}(x+\varepsilon_2 y+\varepsilon_2^2 z) = c_3,$$

and hence the general solution of the PDE is of the form

$$F = \Phi(c_1, c_2, c_3),$$

with Φ of class C^1. At $t = 0$, one imposes

$$\Phi(x + y + z, x + \varepsilon_1 y + \varepsilon_1^2 z, x + \varepsilon_2 y + \varepsilon_2^2 z) = xyz - 1.$$

Denoting

$$u = x + y + z,\ v = x + \varepsilon_1 y + \varepsilon_1^2 z,\ w = x + \varepsilon_2 y + \varepsilon_2^2 z,$$

it follows

$$x = \frac{u+v+w}{3},\ y = \frac{u+\varepsilon_2 v+\varepsilon_1 w}{3},\ z = \frac{u+\varepsilon_1 v+\varepsilon_2 w}{3}.$$

We find $\Phi(u, v, w)$ and then $F(x, y, z, t)$.

Problem 77. *Study the motion of the surface* $S : z^2 + x^2y^2 = 1$ *along the direction of the Euler vector field* (ayz, bzx, cxy).

Problem 78. *The helicoid* $S : x = u\cos v, y = u\sin v, z = bv,\ (u, v) \in \mathbb{R}^2$ *is deformed into*

$$S_t : x = e^t u\cos v, y = e^t u\sin v, z = bv.$$

What happens with the Gauss curvature?

Problem 79. *The helicoid* $S : x = u\cos v, y = u\sin v, z = bv,\ (u, v) \in \mathbb{R}^2$ *is deformed into*

$$S_t : x = u\cos v, y = u\sin v, z = be^{-t}v.$$

What happens with the mean curvature?

Problem 80. *Deform the Tzitzeica surface* $S : xyz = 1$ *by the infinitesimal transformation associated to the normal vector field. Is there another Tzitzeica surface between the deformed ones?*

Problem 81. *We consider the Tzitzeica surface* $S : xyz = 1$ *and the normal vector field* N*. What influence does the flow generated by* N *have on the unit sphere? Between the deformed spheres is there another Tzitzeica surface or not?*

Problem 82. *Let* $S : z = \frac{x^2}{2^2} + \frac{y^2}{3^2}$ *be an elliptic paraboloid. Are there vector fields* X *such that the deformations of* S *through infinitesimal transformations remain convex?*

Problem 83. *Let's consider: (i) the non-algebraic surface*

$$S_1 : F(x, y, z) = -0.4(\sin(5x)+\sin(5y)+\cos(5z))+0.1x^2+0.2y^2+0.2z^2-0.5 = 0;$$

(ii) the tangle (braided cube)

$$S_2 : G(x, y, z) = x^4 - 5x^2 + y^4 - 5y^2 + z^4 - 5z^2 + 10 = 0;$$

(iii) the commode

$$S_3 : H(x, y, z) = (x^2+y^2+z^2-ak^2)^2-b((z-k)^2-2x^2)((z+k)^2-2y^2) = 0,$$

$$\textit{special case: } k = 5, \ a = 0.95 \ \textit{ and } b = 0.8.$$

What they become through the infinitesimal transformation

$$x = x' + t\,x'y'z',\ y = y' + t\,z',\ z = z' + t\,(x' + y')?$$

Problem 84. *Deform the following algebraic surfaces along the normal vector field: Heart,* $(x^2+\frac{9}{4}\,y^2+z^2-1)^3-x^2z^3-\frac{9}{80}\,y^2z^3 = 0$*; Calyx,* $x^2+y^2z^3 = z^4$*; Dingdong,* $x^2+y^2+z^3 = z^2$*; Herz,* $y^2+z^3-z^4-x^2z^2 = 0$*; Sofa,* $x^2 + y^3 + z^5 = 0$*; Zeppelin,* $xyz + yz + 2z^5 = 0$*; Dullo,* $(x^2 + y^2 + z^2)^2 - (x^2 + y^2) = 0$.

Problem 85. *Let us consider the tube (canal) surface*

$$\vec{r}(u, v) = \vec{r}_0(u) + a(\cos v\ \vec{N}(u) + \sin v\vec{B}(u)),$$

where $\vec{N}$ *is the principal normal vector field and* $\vec{B}$ *is the binormal vector field of the spine curve* $v = 0$*, and* a *is the radius of the tube. Find the motion of this surface along the normal vector field.*

Problem 86. *A discrete approximation for the Gaussian function,* $G_\sigma(i, j) = \frac{1}{2\pi\sigma^2}\exp\left[-\frac{i^2+j^2}{2\sigma^2}\right]$*, is called smoothing filter. An image* $I = I(u, v)$ *is smoothed by convolution with a smoothing filter* $G_\sigma(i, j)$*:*

$$S(u, v) = \sum_{i=-n}^{n}\sum_{j=-n}^{n} G_\sigma(i, j)I(u - i, v - j),$$

where the filter kernel is of size $2n + 1$*. Evaluate* $S(u, v)$ *for* $I(u, v) = u^3 + v^3$.

Bibliography

[1] QU Chang-Zheng, LI Yan-Yan, *Higher-dimensional integrable systems arising from motions of curves on $S^2(R)$ and $S^3(R)$*, Commun. Theor. Phys. (Beijing, China) 50, 4 (2008), 841-843.

[2] R. L. Jerrard, D. Smets, *On the motion of a curve by its binormal curvature*, arXiv: 1109.5483v1[math.DG], 26 Sep. 2011.

[3] P. J. Olver, G. Sapiro, A. Tannenbaum, *Invariant geometric evolutions of surfaces and volumetric smoothing*, Manuscript, April 15, 1994.

[4] C. Rogers, W. K. Schief, *Backlund and Darboux Transformations, Geometry and Modern Applications in Soliton Theory*, Cambridge University Press, 2002.

[5] N. Sheridan, *Hamilton's Ricci Flow*, Honours Thesis, The University of Melbourne, Department of Mathematics and Statistics, November 2006.

[6] J. Stam, R. Schmidt, *On the Velocity of an Implicit Surface*, ACM Trans. Graph. XX, Y, Article ZZZ (Month 2010), NN pages. DOI = 10.1145/1559755.1559763; http:doi.acm.org /10.1145/1559755.1559763 ; 2011.

[7] C. Udriste, *Geometric Dynamics*, Mathematics and Its Applications, 513, Kluwer Academic Publishers, Dordrecht, Boston, London, 2000.

[8] C. Udriste, L. Matei, *Lagrange-Hamilton Theories* (in Romanian), Monographs and Textbooks 8, Geometry Balkan Press, Bucharest, 2008.

[9] E. T. Whittaker, *A Treatise on The Analytical Dynamics of Particles & Rigid Bodies*, Cambridge University Press, 1989.

Chapter 9

EVOLUTIVE RIEMANNIAN METRICS

Motto:
"Don't you like green eyes? Here are
Mine! Don't look at them
Too longingly, because you might like them!
Oh, there are tens and hundreds of young people
Who would like to kiss them
But I deny it."

George Coşbuc - *The Lure*

This Chapter studies Riemannian metric flows and waves, including Ricci flow, and Ricci waves. Such mathematical objects are generated by PDEs determined by second order symmetric tensor fields, including the covariant derivative of vector fields with geometric significance.

From the introduction of geometric vector fields on a Riemannian manifold we move to the study of evolutionary PDEs satisfied by the geometric entities induced by the flow of a Riemannian metric. We then present properties of the Ricci flow introduced by R. Hamilton and establish evolutionary PDEs induced by the Riemannian metric dynamics. In context we describe equilibrium points and the fixed states of some geometric evolutions. Finally, the metric flow and the metric wave determined by the Hessian of a function are studied.

The essential ingredients for producing such theories are the flow of the metric and the dynamics of the metric, recently introduced in the specialized literature. This Chapter relates to Richard S. Hamilton's studies [3] in readings about the Ricci flow and more [7], [8].

Physical systems are, in general, dynamical, i.e., they evolve in time. The dynamics of classical mechanical systems are described by Newton's laws of motion, while the dynamics of the classical electromagnetic field is determined by Maxwell's equations. Suggestions for further reading: [1]-[13].

9.1 Geometric vector fields

Let $(M, g(x))$ be a Riemannian manifold of dimension n. The Riemannian metric $g(x) = (g_{ij}(x))$, $i, j, ... = 1, ..., n$, is a symmetric, positive definite, $(0, 2)$-tensor field. The metric $(g_{ij}(x))$ together with its inverse $(g^{ij}(x))$ determine the Christoffel symbols (connection)

$$\Gamma^k_{ij} = \frac{1}{2} g^{k\ell} \left(\frac{\partial g_{\ell i}}{\partial x^j} + \frac{\partial g_{j\ell}}{\partial x^i} - \frac{\partial g_{ij}}{\partial x^\ell} \right)$$

and these determine the curvature tensor field $\mathrm{Riem}(g)$ of components

$$R^\ell{}_{ijk} = \frac{\partial \Gamma^\ell_{ki}}{\partial x^j} - \frac{\partial \Gamma^\ell_{ji}}{\partial x^k} + \Gamma^r_{ki} \Gamma^\ell_{jr} - \Gamma^r_{ji} \Gamma^\ell_{kr}.$$

Taking the trace "$\ell \to k$", it follows the Ricci tensor field $\mathrm{Ric}(g)$ of components $R_{ij} = R^\ell{}_{ij\ell} = g^{\ell k} R_{\ell ijk}$; a new trace with respect to the metric, gives the scalar curvature $R = g^{ij} R_{ij}$.

On a Riemannian manifold there are important PDEs describing vector fields with geometric significance [11], [12], linking the contravariant components X^j or those covariants $X_i = g_{ij} X^j$. These are fixed by the properties of the covariant derivative $\nabla_i X_j = \frac{\partial X_j}{\partial x^i} - \Gamma^h_{ij} X_h$.

A vector field X^j or its dual $X_i = g_{ij} X^j$ is called:

1) *Killing vector field* if and only if its flow consists in isometries. A Killing vector field is a solution of a Killing PDE (covariant form)

$$\nabla_i X_j + \nabla_j X_i = 0 \quad \Leftrightarrow \quad \mathcal{L}_X g = X^k \frac{\partial g_{ij}}{\partial x^k} + g_{kj} \frac{\partial X^k}{\partial x^i} + g_{ik} \frac{\partial X^k}{\partial x^j} = 0.$$

2) *Conformal vector field* if and only if its flow consists in conformal transformations. A conformal vector field is a solution of the conformal PDE

$$\nabla_i X_j + \nabla_j X_i = \psi g_{ij} \quad \Leftrightarrow \quad \mathcal{L}_X g = \psi g,$$

where $n\psi = 2\nabla_i X^i$.

3) *Irrotational vector field* if and only if its flow consists in local irrotational maps. An irrotational vector field is a solution of the irrotational PDE

$$\nabla_i X_j - \nabla_j X_i = 0.$$

4) *Solenoidal vector field* if and only if its flow consists in local solenoidal maps. A solenoidal vector field is a solution of the solenoidal PDE

$$\nabla_i X^i = 0$$

(zero divergence).

5) *Harmonic vector field* if and only if it is irrotational and solenoidal. An harmonic vector field is a solution of the PDEs system

$$\nabla_i X_j - \nabla_j X_i = 0, \quad \nabla_i X^i = 0.$$

We will use these vector fields to generate Riemannian metric flows and Riemannian metric waves.

9.2 Flows of Riemannian metrics

We consider a flow of the Riemannian metric as one parameter family of metrics, $g_{ij}(x,t)$, usually denoted by $g(t)$ or $g_{ij}(t)$ on a fixed Riemannian manifold $(M, g_{ij}(x,0))$ and satisfying a PDE of the form

$$\frac{\partial g_{ij}}{\partial t}(t) = h_{ij}(t), \tag{1}$$

for tensorial field $h_{ij}(t)$ given.

The following result is formulated in the terminology of the paper [2].

Theorem 9.2.1. *Suppose that $g_{ij}(t)$ is a Riemannian metric dependent on time and satisfying the PDE (1). Then the geometric entities attached to the metric $g_{ij}(t)$ evolve through the following PDEs:*

1) *The inverse of the metric:*

$$\frac{\partial g^{ij}}{\partial t} = -h^{ij} = -g^{ik}g^{j\ell}h_{k\ell}.$$

2) *Christoffel symbols:*

$$\frac{\partial \Gamma^k_{ij}}{\partial t} = \frac{1}{2}g^{k\ell}\left(\nabla_i h_{j\ell} + \nabla_j h_{i\ell} - \nabla_\ell h_{ij}\right).$$

The right-hand side member (and hence the left-hand side member) is a tensor.

3) *Riemannian curvature tensor field:*

$$\frac{\partial R^{\ell}_{ijk}}{\partial t} = \frac{1}{2} g^{\ell p} \left(\nabla_i \nabla_j h_{kp} + \nabla_i \nabla_k h_{jp} - \nabla_i \nabla_p h_{jk} \right.$$
$$\left. - \nabla_j \nabla_i h_{kp} - \nabla_j \nabla_k h_{ip} - \nabla_j \nabla_p h_{ik} \right).$$

4) *Ricci tensor field:*

$$\frac{\partial R_{ij}}{\partial t} = \frac{1}{2} g^{pq} \left(\nabla_q \nabla_i h_{jp} + \nabla_q \nabla_j h_{ip} - \nabla_q \nabla_p h_{ij} - \nabla_i \nabla_j h_{qp} \right).$$

5) *Scalar curvature:*

$$\frac{\partial R}{\partial t} = -\Delta H + \nabla^p \nabla^q h_{pq} - h^{pq} R_{pq}, \quad H = g^{pq} h_{pq}.$$

6) *Volume element:*

$$\frac{\partial}{\partial t} d\mu = \frac{H}{2} d\mu.$$

Here $d\mu(x) = \sqrt{\det(g_{ij}(x,t))}\, dx^1 \wedge \ldots \wedge dx^n$.

7) *Volume of a compact manifold:*

$$\frac{d}{dt} \int_M d\mu = \int_M \frac{H}{2} d\mu.$$

8) *Total scalar curvature on a compact manifold M:*

$$\frac{d}{dt} \int_M R d\mu = \int_M \Big(\frac{1}{2} RH - h^{ij} R_{ij} \Big) d\mu.$$

Proof. 1) The evolution of covariant metric $g(t) = (g_{ij}(t))$ induces an evolution of the contravariant metric $g^{-1}(t) = \big(g^{ij}(t)\big)$. Indeed,

$$\frac{\partial}{\partial t} \left(g^{ij} g_{jk} \right) = \frac{\partial}{\partial t} \delta^i_k = 0, \quad \frac{\partial g^{ij}}{\partial t} = -g^{ik} h_{k\ell} g^{\ell j}.$$

2) If the Riemannian metric evolves with respect to the parameter t, then the Riemannian connection evolves with respect to the parameter t. Indeed, for a fixed vector field X, we have

$$\frac{\partial}{\partial t} \nabla_i X = \frac{\partial}{\partial t} \Big(\frac{\partial X^j}{\partial x^i} \frac{\partial}{\partial x^j} + \Gamma^k_{ij} X^j \frac{\partial}{\partial x^k} \Big) = \dot{\Gamma}^k_{ij} X^j \frac{\partial}{\partial x^k}.$$

By the definition of partial derivatives, the derivative $\dot{\Gamma}^k_{ij}$ is a tensor field. To obtain explicitly the tensor $\dot{\Gamma}^k_{ij}$, we use the compatibility with the metric, i.e.,

$$\begin{aligned} 0 = \frac{\partial}{\partial t}\left(\nabla_i g_{jk}\right) &= \frac{\partial}{\partial t}\left(\frac{\partial g_{jk}}{\partial x^i} - \Gamma^\ell_{ij} g_{\ell k} - \Gamma^\ell_{ik} g_{j\ell}\right) \\ &= \nabla_i h_{jk} - \dot{\Gamma}^\ell_{ij} g_{\ell k} - \dot{\Gamma}^\ell_{ik} g_{j\ell}, \end{aligned}$$

and the condition of zero torsion $0 = \dot{\Gamma}^k_{ij} - \dot{\Gamma}^k_{ji}$. Applying cyclic permutations of $\{i, j, k\}$, we can write

$$\begin{aligned} ijk: \quad \nabla_i h_{jk} &= \dot{\Gamma}^\ell_{ij} g_{\ell k} + \dot{\Gamma}^\ell_{ik} g_{j\ell} \\ jki: \quad \nabla_j h_{ki} &= \dot{\Gamma}^\ell_{jk} g_{\ell i} + \dot{\Gamma}^\ell_{ji} g_{k\ell} \\ kij: \quad \nabla_k h_{ij} &= \dot{\Gamma}^\ell_{ki} g_{\ell j} + \dot{\Gamma}^\ell_{kj} g_{i\ell}. \end{aligned}$$

Adding the first two relations and subtracting the third, we find the tensor field

$$\dot{\Gamma}^k_{ij} = \frac{1}{2} g^{k\ell}\left(\nabla_i h_{j\ell} + \nabla_j h_{i\ell} - \nabla_\ell h_{ij}\right).$$

3) Let's highlight the formula for the evolution of the Riemannian curvature tensor field:

$$R^\ell{}_{ijk}\frac{\partial}{\partial x^\ell} = \nabla_i\left(\Gamma^\ell_{jk}\frac{\partial}{\partial x^\ell}\right) - \nabla_j\left(\Gamma^\ell_{ik}\frac{\partial}{\partial x^\ell}\right).$$

The tensor character allows us to calculate $\frac{\partial}{\partial t} R^\ell{}_{ijk} = \dot{R}^\ell{}_{ijk}$ in any advantageous coordinate system. The most appropriate is that of the normal coordinates around a point p, because then Christoffel's symbols are cancelled at the point p. Performing the calculations in this coordinate system, we find

$$\begin{aligned} \dot{R}^\ell{}_{ijk}\frac{\partial}{\partial x^\ell} &= \frac{\partial}{\partial t}\left(\nabla_i\left(\Gamma^\ell_{jk}\frac{\partial}{\partial x^\ell}\right) - \nabla_j\left(\Gamma^\ell_{ik}\frac{\partial}{\partial x^\ell}\right)\right) \\ &= \frac{\partial}{\partial t}\left(\frac{\partial \Gamma^\ell_{jk}}{\partial x^i}\frac{\partial}{\partial x^\ell} - \frac{\partial \Gamma^\ell_{ik}}{\partial x^j}\frac{\partial}{\partial x^\ell}\right) \\ &= \frac{\partial \dot{\Gamma}^\ell_{jk}}{\partial x^i}\frac{\partial}{\partial x^\ell} - \frac{\partial \dot{\Gamma}^\ell_{ik}}{\partial x^j}\frac{\partial}{\partial x^\ell} \\ &= \nabla_i \dot{\Gamma}^\ell_{jk}\frac{\partial}{\partial x^\ell} - \nabla_j \dot{\Gamma}^\ell_{ik}\frac{\partial}{\partial x^\ell} \\ &= \left(\nabla_i \dot{\Gamma}^\ell_{jk} - \nabla_j \dot{\Gamma}^\ell_{ik}\right)\frac{\partial}{\partial x^\ell}, \end{aligned}$$

i.e., $\dot{R}^{\ell}{}_{ijk} = \nabla_i \dot{\Gamma}^{\ell}_{jk} - \nabla_j \dot{\Gamma}^{\ell}_{ik}$. Finally, we get

$$\begin{aligned}
\dot{R}^{\ell}{}_{ijk} &= \nabla_i \left(\frac{1}{2} g^{\ell m} \left(\nabla_j h_{km} + \nabla_k h_{jm} - \nabla_m h_{jk} \right) \right) \\
&\quad - \nabla_j \left(\frac{1}{2} g^{\ell m} \left(\nabla_i h_{km} + \nabla_k h_{im} - \nabla_m h_{ik} \right) \right) \\
&= \frac{1}{2} g^{\ell m} \left(\nabla_i \nabla_j h_{km} - \nabla_j \nabla_i h_{km} + \nabla_i \nabla_k h_{jm} \right. \\
&\quad \left. - \nabla_i \nabla_m h_{jk} - \nabla_j \nabla_k h_{im} + \nabla_j \nabla_m h_{ik} \right) \\
&= \frac{1}{2} g^{\ell m} \left(-R^p_{ijk} h_{mp} - R^p_{ijm} h_{kp} + \nabla_i \nabla_k h_{jm} \right. \\
&\quad \left. - \nabla_i \nabla_m h_{jk} - \nabla_j \nabla_k h_{im} + \nabla_j \nabla_m h_{ik} \right).
\end{aligned}$$

4) Taking the trace $\dot{R}_{jk} = \dot{R}^i{}_{jki}$, we obtain

$$\begin{aligned}
\dot{R}_{jk} &= \frac{1}{2} \Delta_L h_{jk} - \frac{1}{2} g^{im} \nabla_j \nabla_k h_{im} \\
&\quad + \frac{1}{2} g^{im} \left(\nabla_k \nabla_i h_{jm} + \nabla_j \nabla_m h_{ik} \right),
\end{aligned}$$

where

$$\begin{aligned}
\Delta_L h_{jk} &= g^{im} \nabla_i \nabla_m h_{jk} + 2 g^{im} R^p{}_{ijk} h_{mp} \\
&\quad - g^{pq} R_{jq} h_{kp} - g^{pq} R_{kq} h_{jp}
\end{aligned}$$

is the Lichnerowitz Laplacian.

Taking again the trace (with respect to g^{jk}), we find the derivative of scalar curvature

$$\begin{aligned}
\dot{R} &= g^{jk} \dot{R}_{jk} - g^{pj} h_{pq} g^{qk} R_{jk} \\
&= -g^{pq} g^{jk} R_{jq} h_{kp} - g^{im} \nabla_i \nabla_m g^{jk} h_{jk} \\
&\quad + g^{im} g^{jk} \nabla_k \nabla_i h_{jm} \\
&= < Rc, h > - \Delta \mathrm{tr}_g(h) + \mathrm{div}\, \mathrm{div}\, h,
\end{aligned}$$

where $(\mathrm{div}\, h)_j = g^{k\ell} \nabla_k h_{\ell j}$. □

9.3 Ricci flow

In differential geometry, the Ricci flow is an intrinsic geometric flow. It is a process that deforms the metric of a Riemannian manifold in a way formally analogous to the diffusion of heat. Heuristically speaking, at every point of the manifold the Ricci flow shrinks directions of positive curvature and expands directions of negative curvature, while simultaneously smoothing out irregularities in the metric. The latter is analogous to the smoothing behaviour of the heat equation.

Let us consider a Riemannian manifold of dimension $n \geq 2$. Let R_{ij} be the Ricci tensor of the metric variation $g(x,t) = (g_{ij}(x,t))$, with respect to x. In 1982 [3], Richard Hamilton introduced the Ricci flow

$$\frac{\partial g_{ij}}{\partial t}(t) = -2R_{ij}(g(t)),\ g(0) = g^0.$$

In harmonic local coordinates around a point p, the Ricci tensor at p is

$$R_{ij}(p) = -\frac{1}{2}\,\Delta(g_{ij})(p)$$

and so the Ricci flow resembles a heat flow evolution.

If $g^1(t)$ and $g^2(t)$ are Ricci flows on the manifolds M_1 respectively M_2, then $g^1(t) + g^2(t)$ is a Ricci flow on the manifold $M_1 \times M_2$. If $g(t)$ is a Ricci flow, then $\lambda^{-1}g(\lambda t)$ and $g(t+t_0)$ are Ricci flows.

In two dimensions the Ricci flow $\frac{\partial g_{ij}}{\partial t}(t) = -2R_{ij}(g(t))$ is the same as un-normalized Yamabe flow $\frac{\partial g_{ij}}{\partial t}(t) = -Rg_{ij}(t)$. This last PDE is too complicated due to the expression of scalar curvature R with respect to second order derivatives of the metric, regarding x.

Theorem 9.3.1. *([3]) Let (M, g^0_{ij}) be a compact Riemannian manifold. Then there exists a constant $T > 0$ such that the previous Cauchy problem, with given initial values, has a unique solution $g_{ij}(x,t)$ on $M \times [0,T]$.*

The initial idea of Hamilton was to describe the evolution of the metric so that the resulting Riemannian manifold becomes as significant as possible.

Theorem 9.3.2. *([3]) Let M be a 3-dimensional compact manifold which admits a Riemannian metric with strictly positive Ricci curvature. Then M admits a metric with positive constant curvature.*

Particular cases (1) Let's consider the n-dimensional sphere of radius r. The metric is given by $g = r^2\bar{g}$, where $\bar{g}$ is the metric of unit sphere. The sectional curvatures are equal to $\frac{1}{r^2}$. It follows $R_{ij} = \frac{n-1}{r^2}g_{ij} = (n-1)\bar{g}_{ij}$. The Ricci flow is

$$\frac{\partial}{\partial t}(r^2\bar{g}) = -2(n-1)\bar{g}.$$

It rests the PDE $\frac{dr^2}{dt} = -2(n-1)$, with the solution

$$r(t) = \sqrt{R_0^2 - 2(n-1)t},$$

where R_0 is the radius of the initial sphere. The sphere collapses to a point when $t \to \frac{R_0^2}{2(n-1)}$.

(2) For the hyperbolic space (hyperboloid model in $\mathbb{R}^{n+1}$)

$$H^n : (x^1)^2 + ... + (x^n)^2 - (x^{n+1})^2 = -1, \;\; x^{n+1} > 0,$$

the Ricci flow is reduced to the ODE $\frac{dr^2}{dt} = 2(n-1)$, with the general solution $r(t) = \sqrt{R_0^2 + 2(n-1)t}$. Obviously $\lim_{t\to\infty} r(t) = \infty$.

(3) If $R_{ij} = \frac{R}{n} g_{ij}$, $R =$ constant (scalar curvature), then the manifold is Einstein manifold. It follows

$$\frac{\partial g_{ij}}{\partial t}(t) = -2\,\frac{R}{n}\, g_{ij}(t), \;\; g(0) = g^0,$$

with the solution $g_{ij} = g_{ij}^0\, e^{\frac{-2}{n}Rt}$.

(4) Equilibrium points: A Riemannian manifold is flat, i.e., locally equivalent to the Euclidean space, if and only if the curvature Riemannian tensor is zero. Then $R_{ij} = 0$ and the corresponding metric $g(x, 0) = g^0(x)$ is an equilibrium point.

(5) Ricci solitons: Let $g(t)$ be a solution of the Ricci flow, let $\varphi_t : M \to M$, $\varphi_0 = id$, be a diffeomorphism dependent on time and let $\sigma(t)$, $\sigma(0) = 1$ be a scaling factor. If $g(t) = \sigma(t)\varphi_t^*(t)g(0)$, i.e.,

$$g_{ij}(t) = \sigma(t)\frac{\partial \varphi_t^{*i'}}{\partial x^i}\frac{\partial \varphi_t^{*j'}}{\partial x^j} g_{i'j'}(0),$$

then the solution $g(t)$ is called *Ricci soliton*. We compute the partial derivative

$$\frac{\partial}{\partial t} g(t) = \frac{d\sigma}{dt}\varphi_t^*(t)g(0) + \sigma(t)\frac{\partial}{\partial t}\varphi_t^*(t)g(0)$$

and then we set $t = 0$. It follows

$$-2R_{ij}(g(0)) = \sigma'(0)g_{ij}(0) + \mathcal{L}_V g_{ij}(0).$$

We denote $\sigma'(0) = 2\lambda$, $V = \frac{d\varphi_t}{dt}$ and $V_i = \nabla_i f$. We find the *Ricci soliton of gradient type PDE*

$$R_{ij} + 2\lambda g_{ij} + \nabla_i\nabla_j f = 0,$$

whose unknown is the metric $g_{ij}(x)$. A Ricci soliton of gradient type is called: (1) *contractive* if $\lambda < 0$; (2) *expansive* if $\lambda > 0$; (3) *static* if $\lambda = 0$.

Theme (Hamilton's cigar soliton) Show that $g_{ij} = \frac{1}{1+x^2+y^2}\delta_{ij}$ is a soliton of gradient type on $\mathbb{R}^2$, with respect to the function $f(x, y) = \frac{1}{2}(x^2 + y^2)$.

9.4 Riemannian metric dynamics

The wave equation is an important second-order PDE. This kind of PDE is encountered in classical physics - such as mechanical waves (e.g. water waves, sound waves and seismic waves) or light waves. It arises in fields like acoustics, electromagnetics, and fluid dynamics.

We start with a fixed Riemannian manifold $(M, g_{ij}(x, 0))$. Let us study here the Riemannian metric wave (as a Riemannian metric family with a parameter t), usually denoted by $g(t) = (g_{ij}(t))$, with meaning $g(x, t) = (g_{ij}(x, t))$, satisfying a PDE of the form

$$\frac{\partial^2 g_{ij}}{\partial t^2}(t) = k_{ij}(t), \tag{2}$$

for a prescribed tensor field $k_{ij}(t)$.

Theorem 9.4.1. *Suppose that $g_{ij}(t)$ is a Riemann metric depending on time. Then the fundamental geometric entities attached to the metric evolve according to the following PDEs:*

1) *Inverse of the metric:*

$$\frac{\partial^2 g^{i\ell}}{\partial t^2} = -\Big(2\frac{\partial g^{ij}}{\partial t}h_{jk} + g^{ij}k_{jk}\Big)g^{k\ell}.$$

2) *Christoffel symbols:*

$$\frac{\partial^2 \Gamma^\ell_{ij}}{\partial t^2} = \frac{1}{2}g^{\ell k}\left(\nabla_i k_{jk} + \nabla_j k_{ki} - \nabla_k k_{ij}\right) - 2\dot{\Gamma}^m_{ij} g^{\ell k} h_{km}.$$

Proof. 1) On the one hand, the flow evolution of the covariant metric $g(t) = (g_{ij}(t))$ induces a flow of contravariant metric $g^{-1}(t) = \big(g^{ij}(t)\big)$. Indeed,

$$\frac{\partial}{\partial t}\left(g^{ij}g_{jk}\right) = \frac{\partial}{\partial t}\delta^i_k = 0, \quad \frac{\partial g^{ij}}{\partial t} = -g^{ik}h_{k\ell}.$$

On the other hand, the wave evolution of the covariant metric $g(t) = (g_{ij}(t))$ generates a wave evolution of the contravariant metric $g^{-1}(t) = \big(g^{ij}(t)\big)$. To show that, we use

$$\frac{\partial^2}{\partial t^2}\left(g^{ij} g_{jk}\right) = \frac{\partial^2}{\partial t^2}\delta^i_k = 0,$$

$$\frac{\partial}{\partial t}\left(\frac{\partial g^{ij}}{\partial t} g_{jk} + g^{ij} h_{jk}\right) = 0.$$

It follows

$$\frac{\partial^2 g^{ij}}{\partial t^2} g_{jk} + \frac{\partial g^{ij}}{\partial t} h_{jk} + \frac{\partial g^{ij}}{\partial t} h_{jk} + g^{ij} k_{jk} = 0,$$

or

$$\frac{\partial^2 g^{i\ell}}{\partial t^2} = -\Big(2\frac{\partial g^{ij}}{\partial t} h_{jk} + g^{ij} k_{jk}\Big) g^{k\ell}.$$

2) In order to find the wave type evolution of the Riemannian connection, we start with

$$\begin{aligned}\frac{\partial^2}{\partial t^2}\nabla_i X &= \frac{\partial^2}{\partial t^2}\Big(\frac{\partial X^j}{\partial x^i}\frac{\partial}{x^j} + \Gamma^k_{ij} X^j \frac{\partial}{x^k}\Big)\\ &= X^j \ddot{\Gamma}^k_{ij}\frac{\partial}{\partial x^k}.\end{aligned}$$

To obtain explicitly the tensor $\ddot{\Gamma}^k_{ij}$, we use the compatibility with the metric

$$\begin{aligned}0 &= \frac{\partial^2}{\partial t^2}\left(\nabla_i g_{jk}\right)\\ &= \frac{\partial}{\partial t^2}\Big(\frac{\partial g_{jk}}{\partial x^i} - \Gamma^\ell_{ij} g_{\ell k} - \Gamma^\ell_{ik} g_{j\ell}\Big)\\ &= \frac{\partial}{\partial t}\big(\nabla_i \dot{g}_{jk} - \dot{\Gamma}^\ell_{ij} g_{\ell k} - \dot{\Gamma}^\ell_{ik} g_{j\ell}\big)\\ &= \frac{\partial}{\partial t}\Big(\frac{\partial h_{jk}}{\partial x^i} - \Gamma^\ell_{ij} h_{\ell k} - \Gamma^\ell_{ik} h_{j\ell} - \dot{\Gamma}^\ell_{ij} g_{\ell k} - \dot{\Gamma}^\ell_{ik} g_{j\ell}\Big)\\ &= \frac{\partial k_{jk}}{\partial x^i} - \dot{\Gamma}^\ell_{ij} h_{\ell k} - \Gamma^\ell_{ij} k_{\ell k} - \dot{\Gamma}^\ell_{ik} h_{j\ell} - \Gamma^\ell_{ik} k_{j\ell}\\ &\quad -\ddot{\Gamma}^\ell_{ij} g_{\ell k} - \dot{\Gamma}^\ell_{ij} h_{\ell k} - \ddot{\Gamma}^\ell_{ik} g_{j\ell} - \dot{\Gamma}^\ell_{ik} h_{j\ell}\\ &= \nabla_i k_{jk} - 2\dot{\Gamma}^\ell_{ij} h_{\ell k} - 2\dot{\Gamma}^\ell_{ik} h_{j\ell} - \ddot{\Gamma}^\ell_{ij} g_{\ell k} - \ddot{\Gamma}^\ell_{ik} g_{j\ell}\end{aligned}$$

and $0 = \dot{\Gamma}^k_{ij} - \dot{\Gamma}^k_{ji}$ (see, free of torsion). Applying the cyclique permutations of the set $\{i, j, k\}$, we find

$$ijk : \nabla_i k_{jk} = 2\dot{\Gamma}^\ell_{ij} h_{\ell k} + 2\dot{\Gamma}^\ell_{ik} h_{j\ell} + \ddot{\Gamma}^\ell_{ij} g_{\ell k} + \ddot{\Gamma}^\ell_{ik} g_{j\ell}$$
$$jki : \nabla_j k_{ki} = 2\dot{\Gamma}^\ell_{jk} h_{\ell i} + 2\dot{\Gamma}^\ell_{ji} h_{k\ell} + \ddot{\Gamma}^\ell_{jk} g_{\ell i} + \ddot{\Gamma}^\ell_{ji} g_{k\ell}$$
$$kij : \nabla_k k_{ij} = 2\dot{\Gamma}^\ell_{ki} h_{\ell j} + 2\dot{\Gamma}^\ell_{kj} h_{i\ell} + \ddot{\Gamma}^\ell_{ki} g_{\ell j} + \ddot{\Gamma}^\ell_{kj} g_{i\ell}.$$

Adding the first two relations and subtracting the third, we deduce

$$\nabla_i k_{jk} + \nabla_j k_{ki} - \nabla_k k_{ij} = 4\dot{\Gamma}^\ell_{ij} h_{k\ell} + 2\ddot{\Gamma}^\ell_{ij} g_{\ell k}$$

i.e.,

$$\ddot{\Gamma}^\ell_{ij} = \frac{1}{2} g^{\ell k} \left(\nabla_i k_{jk} + \nabla_j k_{ki} - \nabla_k k_{ij} \right) - 2\dot{\Gamma}^m_{ij} g^{\ell k} h_{km},$$

which completes the proof. □

Open problem Study the wave evolution of Riemannian curvature tensor, of Ricci tensor, and of scalar curvature. Some results are found in [5].

9.5 Ricci wave

Let us consider a Riemannian manifold of dimension $n \geq 2$. In 2007, Kong and Liu [6] introduced the Ricci wave (under the name hyperbolic geometric flow), using the Cauchy problem (PDE of evolution)

$$\frac{\partial^2 g_{ij}}{\partial t^2}(t) = -2R_{ij}(g(t)),\ g(0) = g^0,\ \frac{\partial g}{\partial t}(0) = h^0,$$

where R_{ij} is the Ricci tensor of the metric $g(x, t) = (g_{ij}(x, t))$, with respect to x. Their idea was to describe the wave character of the metric so that the resulting Riemannian manifold becomes as significant as possible.

Theorem 9.5.1 (Kong, Liu, 2007). *Let (M, g^0_{ij}) be a compact Riemannian manifold. Then there exists a constant $T > 0$ such that the previous Cauchy problem has a unique solution $g_{ij}(x, t)$ on $M \times [0, T]$.*

If the metric $g_{ij}(x, 0)$ is Ricci flat, i.e., $R_{ij}(g(x, 0)) = 0$, then $g_{ij}(x, t) = g_{ij}(x, 0)$ is an equilibrium state (fixed).

If $R_{ij} = \frac{R}{n} g_{ij}$, $R =$ constant (scalar curvature), then the manifold is Einstein manifold. It follows

$$\frac{\partial^2 g_{ij}}{\partial t^2}(t) = -2\,\frac{R}{n}\,g_{ij},$$

with the solution (for $R > 0$)

$$g_{ij} = g^0_{ij}\,\cos\sqrt{\frac{2R}{n}}\;t + h^0_{ij}\,\sin\sqrt{\frac{2R}{n}}\;t.$$

Suppose that the initial metric $g_{ij}(x,0)$ is Einstein, i.e., $R_{ij}(g(x,0)) = \lambda g_{ij}(x,0)$. Let us set $g_{ij}(x,t) = \rho(t) g_{ij}(x,0)$. By the definition of Ricci tensor, we find $R_{ij}(g(x,t)) = R_{ij}(g(x,0)) = \lambda g_{ij}(x,0)$. The Ricci wave PDE becomes

$$\frac{\partial^2}{\partial t^2}\rho(t) g_{ij}(x,0) = -2\lambda g_{ij}(x,0),$$

i.e., the ODE $\frac{d^2\rho}{dt^2}(t) = -2\lambda$. Setting the initial conditions $\rho(0) = 1$, $\rho'(0) = v$, we find the solution $\rho(t) = -\lambda t^2 + vt + 1$.

A cosmological example A typical example of the Einstein metric is

$$ds^2 = \frac{1}{1-\kappa r^2} dr^2 + r^2 d\theta^2 + r^2 \sin^2\theta\, d\varphi^2,$$

where κ is a constant taking the values $-1, 0$ or 1. We observe that

$$ds^2 = R^2(t)\left(\frac{1}{1-\kappa r^2} dr^2 + r^2 d\theta^2 + r^2 \sin^2\theta\, d\varphi^2\right)$$

is a solution of the previous wave PDE if

$$R^2(t) = -2\kappa t^2 + c_1 t + c_2,$$

where c_1 and c_2 are two arbitrary constants.

9.6 Flows of metrics determined by geometric vector fields

We consider a flow of the Riemannian metric as one parameter family of metrics, $g_{ij}(x,t)$, usually denoted by $g(t)$ or $g_{ij}(t)$ on a fixed Riemannian manifold $(M, g_{ij}(x,0))$ and satisfying an evolution PDE.

1) Let ξ_i be a vector field of class C^1 that determines the flow $\frac{\partial g_{ij}}{\partial t} = \nabla_i \xi_j$. Since the metric g_{ij} is symmetric, it follows that the covariant derivative $\nabla_i \xi_j$ satisfies $\nabla_i \xi_j = \nabla_j \xi_i$, i.e., ξ must be an irrotational vector field.

Proposition 9.6.1. *If $\xi(x)$ is a parallel vector field with respect to the metric $g(x)$, then the metric $g(t,x) = g(x)$ is an equilibrium point of the previous flow.*

2) We introduce the flow $\frac{\partial g_{ij}}{\partial t} = \frac{1}{2}\left(\nabla_i \xi_j + \nabla_j \xi_i\right)$, using the symmetric part of the covariant derivative $\nabla_i \xi_j$.

Proposition 9.6.2. *If $\xi(x)$ is a Killing vector field with respect to the metric $g(x)$, then the metric $g(x,t) = g(x)$ is an equilibrium point of the previous flow.*

Theorem 9.6.1. *If $\xi(x)$ is a conformal vector field with respect to the metric $h(x)$ and if $\operatorname{div}\xi > 0$, then the metric $g(x,t) = \frac{1}{2}\psi(x)h(x)t + a(x)$, $t \in (0,\infty)$ is an equilibrium point of the flow.*

Proof. It is sufficient to remark that the PDE $\frac{\partial g}{\partial t}(t,x) = \frac{1}{2}\psi(x)h(x)$ has a solution of the form $g(x,t) = \frac{1}{2}\psi(x)h(x)t + a(x)$. □

9.7 Dynamics of metrics determined by geometric vector fields

1) Let ξ_i be a vector field whose covariant derivative $\nabla_i \xi_j$ determines the wave $\frac{\partial^2 g_{ij}}{\partial t^2} = \nabla_j \xi_i$. By the symmetry of the metric, it follows that ξ_i must be an irrotational vector field.

Theorem 9.7.1. *If $\xi(x)$ is a parallel vector field with respect to the metric $g_{ij}(x)$, then the metric $g_{ij}(x,t) = a_{ij}(x)t + b_{ij}(x)$, $t \in (0,\infty)$ is a fixed state for the previous wave.*

2) Let us introduce the wave

$$\frac{\partial^2 g_{ij}}{\partial t^2} = \frac{1}{2}\left(\nabla_i \xi_j + \nabla_j \xi_i\right),$$

determined by the symmetric part of the covariant derivative $\nabla_i \xi_j$.

Theorem 9.7.2. *If $\xi(x)$ is a Killing vector field with respect to the metric $g_{ij}(x)$, then the metric $g_{ij}(x,t) = a_{ij}(x)t + b_{ij}(x)$, $t \in (0,\infty)$ is a fixed state for the previous wave PDE.*

Theorem 9.7.3. *If $\xi(x)$ is a conformal vector field with respect to the metric $h(x)$ and if* $\operatorname{div}\xi > 0$ *(evolutive vector field), then the metric $g(x,t) = \frac{1}{4}\psi(x)h(x)t^2 + a(x)t + b(x)$, with $t \in (0,\infty)$ is a fixed state of the previous wave PDE.*

Proof. From $\frac{\partial^2 g}{\partial t^2}(t,x) = \frac{\psi(x)}{2}h(x)$, it follows that $\frac{\partial g}{\partial t} = \frac{\psi(x)}{2}h(x)t + a(x)$. Integrating again, we deduce $g(x,t) = \frac{1}{4}\psi(x)h(x)t^2 + a(x)t + b(x)$. □

9.8 Metric flow determined by a Hessian

Let $(M, g(x,t))$ be a Riemannian manifold and $f : M \to \mathbb{R}$ be a C^2 function. We introduce the Hessian of f with respect to g, denoted by $\mathrm{Hess}_{g(x,t)} f(x)$, by the components

$$\frac{\partial^2 f}{\partial x^i \partial x^j} - \Gamma^k_{ij}\frac{\partial f}{\partial x^k}$$

(second order covariant derivative).

The previous Hessian determines the flow (see also [9], [10])

$$\frac{\partial g}{\partial t}(x,t) = \mathrm{Hess}_{g(x,t)} f(x). \tag{3}$$

If $f(x)$ is geodesic linear affine with respect to the metric $g(x)$, i.e.,

$$\mathrm{Hess}_{g(x,0)} f(x) = 0,$$

then the initial metric $g(x,0) = g(x)$ is an equilibrium point of the flow.

Let $U \subset M$ be an open connected subset (domain), with compact closure. The volume of U is

$$V_g(U) = \int_U \sqrt{\det g}\, dx^1 \wedge \ldots \wedge dx^n.$$

If the Riemannian metric depends differentiable on one real parameter t, then

$$V = V_g(U;t) = \int_U \sqrt{\det g(t)}\, dx^1 \wedge \ldots \wedge dx^n.$$

and

$$\begin{aligned}\frac{dV}{dt} &= \frac{1}{2}\int_U \sqrt{\det g(t)}\, g^{ij}\dot{g}_{ij}\, dx^1 \wedge \ldots \wedge dx^n \\ &= \frac{1}{2}\int_U \sqrt{\det g(t)}\, \mathrm{tr}_g \dot{g}\, dx^1 \wedge \ldots \wedge dx^n.\end{aligned}$$

Theorem 9.8.1. *On a compact manifold M, endowed with a Riemannian metric $g(x,t)$, which is the solution of the evolution PDE (3), the volume does not depend on the parameter t.*

Proof. Applying the formula

$$\frac{dV}{dt} = \frac{1}{2}\int_M \sqrt{\det g}\, g^{ij}\,\dot{g}_{ij}\, dx^1 \wedge \ldots \wedge dx^n$$

for the metric $g(x,t)$ which satisfies the evolution PDE (3), we obtain

$$\begin{aligned}\frac{dV}{dt} &= \frac{1}{2}\int_M \sqrt{\det g}\, g^{ij}\,(\nabla_i\nabla_j f)\, dx^1 \wedge \ldots \wedge dx^n \\ &= \frac{1}{2}\int_M \Delta f\,\sqrt{\det g}\; dx^1 \wedge \ldots \wedge dx^n.\end{aligned}$$

Taking into account the compacity of the manifold M and applying Green Theorem, we obtain $\frac{dV}{dt} = 0$. Therefore, the volume V is constant. □

On the other hand, on non-compact Riemmanian manifolds there exist non-constant convex functions in Riemannian sense. In this context, it follows [10]

Theorem 9.8.2. *Let $(M, g(x,t))$ be a Riemannian manifold with respect to the metric $g(x,t)$ that satisfies the evolution* PDE $\frac{\partial g}{\partial t} = \mathrm{Hess}_g f$. *If f is a non-constant convex function, with respect to the metric $g(x,t)$, then the volume of any domain in M is increasing with respect to t.*

Proof. The function f is convex with respect to g if and only if the Hessian $\mathrm{Hess}_g\, f$ is positive semi-definite. Consequently, $\Delta f = g^{-1}(\mathrm{Hess}_g f) \geq 0$ and $\frac{dV}{dt} \geq 0$. Hence the function $V(t)$ is increasing with respect to t. □

9.9 Metric dynamics determined by a Hessian

Let $(M, g(x,t))$ be a Riemannian manifold of dimension n, and $f : M \to \mathbb{R}$ be a C^2 function. The Hessian of the function f determines the wave PDE

$$\frac{\partial^2 g}{\partial t^2}(x,t) = \mathrm{Hess}_{g(x,t)} f(x). \tag{4}$$

If the function $f(x)$ is geodesic linear affine with respect to the metric $g(x)$, then the initial metric $g(x,0) = g(x)$ is a fixed state.

Let $U \in M$ be a connected open subset (domain) with compact closure. If we denote $dv = \sqrt{\det g(t)}\; dx^1 \wedge \ldots \wedge dx^n$, then the volume V of U satisfies

$$\frac{dV}{dt} = \frac{1}{2}\int_U \operatorname{tr}_g \dot{g}\, dv$$

$$\frac{d^2V}{dt^2} = \frac{1}{2}\int_U \left(\frac{1}{2}\left(g^{ij}\dot{g}_{ij}\right)^2 + \dot{g}^{ij}\dot{g}_{ij} + g^{ij}\ddot{g}_{ij}\right) dv.$$

Since $\dot{g}^{ij} = -g^{ik}\dot{g}_{k\ell}g^{\ell j}$, we find

$$\frac{d^2V}{dt^2} = \frac{1}{2}\int_U \left(\frac{1}{2}\left(\operatorname{tr}_g\dot{g}\right)^2 - \|\dot{g}\|^2 + \Delta f\right) dv.$$

Consequently, we have

Theorem 9.9.1. *Let M be a compact manifold endowed with the Riemannian metric $g(x,t)$, which is a solution of the evolution PDE* (4).

1) If there exist the real numbers a and b such that

$$a\left(\frac{1}{2}(\operatorname{tr}_g\dot{g})^2 - ||\dot{g}||^2\right) + b\,(\operatorname{tr}_g\dot{g}) = 1,$$

then the volume function V is a solution of ODE (see harmonic oscillator with damping)

$$a\frac{d^2V}{dt^2} + b\frac{dV}{dt} - \frac{V}{2} = 0.$$

2) If $\max_{t,x}\|\dot{g}\| = \lambda$, $\max_{t,x}\operatorname{tr}_g\dot{g} = \mu$, then the volume function V satisfies the differential inequalities

$$-\frac{\lambda^2}{2}V \leq \frac{d^2V}{dt^2} \leq \frac{\mu^2}{4}V.$$

Proof. The previous relations are consequences of the formulas

$$\frac{dV}{dt} = \frac{1}{2}\int_M \operatorname{tr}_g\dot{g}\; dv,$$

$$\frac{d^2V}{dt^2} = \frac{1}{2}\int_M \left(\frac{1}{2}\left(\operatorname{tr}_g\dot{g}\right)^2 - \|\dot{g}\|^2\right) dv.$$

□

9.10 Self-evaluation problems

Problem 87. *Let us consider the special case $n = 1$, i.e., the initial Riemannian manifold is $(\mathbb{R}, g_0(x))$. For a given non-constant C^2 function $f : \mathbb{R} \to \mathbb{R}$, solve the Cauchy problem*

$$\frac{\partial g}{\partial t}(x,t) = f''(x) - \frac{f'(x)}{2g(x,t)}\frac{\partial g}{\partial x}(x,t),\ g(x,0) = g_0(x),$$

whose unknown is a Riemannian metric variation $g(x,t)$.

Solution. To solve this Cauchy problem, we use the characteristic system

$$\frac{dt}{1} = \frac{dx}{\frac{f'(x)}{2g(x,t)}} = \frac{dg}{f''(x)},$$

which has two first integrals

$$\frac{f'(x)}{\sqrt{g(x,t)}} = c_1,\ g(x,t) - f''(x)t = c_2.$$

This Cauchy problem has a unique solution $g(x,t)$, if f and g_0 do not satisfy the characteristic equations. Therefore the general solution $g(x,t)$ of the PDE is the solution of the equation

$$\frac{f'(x)}{\sqrt{g(x,t)}} = \Phi\left(g(x,t) - tf''(x)\right),$$

where Φ is an arbitrary function of class C^1. For $t = 0$, we find

$$\frac{f'(x)}{\sqrt{g_0(x)}} = \Phi(g_0(x)).$$

Setting $u = g_0(x)$, and supposing that $g_0(x)$ is invertible, we get $\Phi(u) = \frac{f'(g_0^{-1}(u))}{\sqrt{u}}$. Coming back, we find the implicit equation

$$\frac{f'(x)}{\sqrt{g(x,t)}} = \frac{f'(g_0^{-1}(g(x,t) - tf''(x)))}{\sqrt{g(x,t) - tf''(x)}}.$$

Problem 88. *On the Riemannian manifold $(\mathbb{R}^2, g_{ij}(x,y))$, we introduce the variation $g_{ij}(x,y,t)$, where*

$$g_{11}(x,y,t) = f(x)c_1(t),\ g_{22}(x,y,t) = f(x)c_2(t),\ g_{12}(x,y,t) = 0.$$

Find the non-zero components of the Ricci tensor field. Applying Maple, we get

$$R_{11} = \frac{1}{2}\frac{d^2}{dx^2}\ln f(x),\ R_{22} = \frac{1}{2}\frac{c_2(t)}{c_1(t)}\frac{d^2}{dx^2}\ln f(x).$$

Find $f(x)$ such that all components of the Ricci tensor are zero. Answer: $f(x) = C_2 e^{C_1 x}$. Find $c_1(t)$ and $c_2(t)$, via Ricci flow PDEs.

Problem 89. *Suppose $n = 1$, i.e., the initial Riemannian manifold is $(\mathbb{R}, g_0(x))$. For a given non-constant C^2 function $f : \mathbb{R} \to \mathbb{R}$, we have the Cauchy problem*

$$\frac{\partial^2 g}{\partial t^2}(x,t) = f''(x) - \frac{f'(x)}{2g(x,t)}\frac{\partial g}{\partial x}(x,t),\ g(x,0) = g_0(x),\ \frac{\partial g}{\partial t}(x,0) = h(x),$$

whose unknown is the Riemannian metric variation $g(x,t)$. Discuss the solution of this Cauchy problem.

Problem 90. *Let R_g be the scalar curvature of the compact Riemannian manifold (M, g). The Perelman's functional is*

$$\mathcal{F}(g_{ij}(\cdot), f(\cdot)) = \int_M (R_g + \|\nabla f\|^2)\exp(-f)dV_g,$$

where dV_g represents the volume element of the manifold M with respect to the metric g. Find the Euler-Lagrange PDEs.

Problem 91. *Let $M = \mathbb{R} \times S^1$, where S^1 is the circle of radius 1. We define a solution*

$$g(x,\theta,t) = \frac{\sinh t}{\cosh x + \cosh t}(dx^2 + d\theta^2),\ t > 0$$

to the Ricci flow on M. Show that the the scalar curvature of this metric is given by

$$R(x,\theta,t) = \frac{1 + \cosh t \cosh x}{\sinh t(\cosh x + \cosh t)} > 0,$$

and attains its maximum curvature at the poles, i.e., for $x \to \pm\infty$:

$$R_{\max}(t) = \lim_{|x|\to\infty} R(x,\theta,t) = \coth t > 0,\ \forall t > 0.$$

Problem 92. *We consider the Riemannian manifold $(\mathbb{R}^2, g_{ij}(x,y))$, with $g_{12} = 0$. Compute the Ricci tensor field using Maple facilities. Answer: Denote*

$$A = 2g_{11}g_{22}\partial^2_{y^2}g_{11} + 2g_{11}g_{22}\partial^2_{x^2}g_{22} - g_{11}\partial_y g_{22}\partial_y g_{11}$$

$$-g_{11}(\partial_x g_{22})^2 - g_{22}\partial_x g_{22}\partial_x g_{11} - g_{22}(\partial_y g_{11})^2.$$

Then

$$R_{11} = \frac{1}{4}\frac{1}{g_{11}(g_{22})^2}\,A,\ R_{22} = \frac{1}{4}\frac{1}{(g_{11})^2 g_{22}}\,A.$$

Problem 93. *Let $S(q,n)$ be the set of real symmetric quadratic matrices, of rank q and order n. The Riccati flow*

$$\dot{X} = (A - X)X + X(A - X),\ X(0) = X_0 \in S(q,n)$$

keeps the rank on $S(q,n)$. If the matrix A is invertible, show that the solution of this Cauchy problem is

$$X(t) = e^{tA}X_0[I_n + A^{-1}(e^{2tA} - I_n)X_0]^{-1}e^{tA}.$$

Furthermore, for each $X_0 \geq 0$ (positive semidefinite), the solution $X(t)$, $t \in [0,\infty)$ exists and is positive semidefinite. Reformulate the problem in Riemannian metric language.

Hint *See [4] pg. 141.*

Problem 94. *Show that the dynamical system*

$$\dot{X} = X^2(A - X) + (A - X)X^2,$$

defined on the vector space of symmetric matrices, preserves the rank, i.e., $\operatorname{rank} X(t) = \operatorname{rank} X(0)$, $\forall t \geq 0$, and consequently keeps the signature. The matrix solution $X(t)$ has the limit $\lim_{t\to\infty} X(t) = X(\infty)$, where the matrix $X(\infty)$ is also a symmetric, and so $\operatorname{rank} X(\infty) \leq \operatorname{rank} X(0)$. Reformulate the problem in Riemannian metric language.

Problem 95. *Let $I \subset \mathbb{R}$ be an interval and $A(t) \in S(q,n)$, $t \in I$ be a continuous family of matrices. Show that*

$$\dot{X}(t) = A(t)X(t) + X(t)A(t)^T,\ X(0) \in S(q,n)$$

is a flow on $S(q,n)$ which preserves the rank and hence the signature. Conversely, any flow on $S(q,n)$ which preserves the rank is of this form. Reformulate the problem in Riemannian metric language.

Problem 96. *On the Riemannian manifold* $(\mathbb{R}, g(t,x))$, *analyse the metric flow, determined by the Hessian of a* C^2 *function* f, *i.e., the PDE*

$$\frac{\partial g}{\partial t}(t,x) = \frac{d^2 f}{dx^2}(x) - \frac{1}{2g(t,x)}\frac{\partial g}{\partial x}(t,x)\frac{df}{dx}(x).$$

Details for the functions $f(x) = x^2$, *respectively* $f(x) = x^3$.

Problem 97. *On the Riemannian manifold* $(\mathbb{R}, g(t,x))$, *investigate the dynamics (wave) of the metric, determined by a Hessian of a* C^2 *function* f, *i.e., the PDE*

$$\frac{\partial^2 g}{\partial t^2}(t,x) = \frac{d^2 f}{dx^2}(x) - \frac{1}{2g(t,x)}\frac{\partial g}{\partial x}(t,x)\frac{df}{dx}(x).$$

Details for the functions $f(x) = x^2$, *respectively* $f(x) = x^3$.

Problem 98. *Study the Calabi flow* $\frac{\partial g_{ij}}{\partial t}(x,t) = (\Delta_x R) g_{ij}(x,t)$ *on the Riemannian manifold* $(\mathbb{R}^2, g_{ij}(x,0) = g^0_{ij}(x))$.

Bibliography

[1] C. Corcodel, C. Udriste, *Evolution metrics and geometric vector fields*, ICOSSSE-10 Proceedings of the 9th WSEAS International Conference on System Science and Simulation in Engineering, 2010, 59-64.

[2] D. Glickenstein, *Introduction to Flows on Riemannian Metrics*, Math. 538, Springer, 2009.

[3] R. S. Hamilton, *Three-manifolds with positive Ricci curvature*, J. Differ. Geom., 17, 2 (1982), 255-306.

[4] U. Helmke, J. B. Moore, *Optimization and Dynamical Systems*, 2nd Edition, University of Würzburg, Germany; Australian National University, Canberra, Australia; March 1996.

[5] I. E. Hirica, C. Udriste, *Basic evolution in Riemannian geometry*, Balk. J. Geom. Appl., 17, 1 (2012), 30-40.

[6] D. -X. Kong, K. Liu, *Wave character of metrics and hyperbolic geometry*, http://www.cms.zju.edu.cn/UploadFiles/Attach Files/200682885946597.pdf

[7] J. Morgan, G. Tian, *Ricci Flow and the Poincaré Conjecture*, Clay Math. Monographs, Vol. 3, Amer. Math. Soc. and Clay Math. Institute, 2007.

[8] N. Sheridan, *Hamilton's Ricci Flow*, Honours Thesis, The University of Melbourne, Department of Mathematics and Statistics, November 2006.

[9] C. Udriste, M. Toda, *Volume dependence on the Riemannian metric*, Sci. Bull. PUB, Series A: Appl. Math. Phys. 55, 3-4 (1993), 285-298.

[10] C. Udriste, M. Toda, *Influence of the evolution Riemannian metrics on the volume*, in "Proc. 23rd Conf. Geom. Topol., D. Andrica et al. (Eds.), Cluj-Napoca, 1994, pp. 194-202.

[11] C. Udriste, *Field Lines* (in Romanian), Technical Publishing House, Bucharest, 1988.

[12] C. Udriste, *Geometric Dynamics*, Mathematics and Its Applications, 513, Kluwer Academic Publishers, Dordrecht, Boston, London, 2000.

[13] C. Udriste, L. Matei, *Lagrange-Hamilton Theories* (in Romanian), Monographs and Textbooks 8, Geometry Balkan Press, Bucharest, 2008.

Chapter 10

EVOLUTION OF TZITZEICA HYPERSURFACES

Motto:
"Why this hurly-burly
Just some boisterous kids
Hey, indeed! May I live to see them all grown-ups
I thought it was a troop
of Huns!"
George Coşbuc - *Winter in the Village-Street*

In this Chapter, our aim is to study the evolutions of Tzitzeica hypersurfaces which appear in understanding the dynamics of some geometric programming problems and reliability optimal allocation problems. Section 2 *analyses the convexity of a Tzitzeica hypersurface. Sections* $3-6$ *refer to standard Tzitzeica hypersurfaces and their evolutions by convenient geometrical flows: (i) evolution along the normal vector field, (ii) infinitesimal normal transformation of Tzitzeica hypersurface, (iii) evolution along a centro-affine vector field, and (iv) evolution along an affine vector field. Sections* $7-8$ *include results on Tzitzeica law in economics and evolution of Tzitzeica surfaces described by PDEs: (1) Tzitzeica hypersurfaces as invariants with respect to excess demand flow; (2) parametric Tzitzeica surfaces based on PDEs and their evolutions. Suggestions for further reading: [1]-[10].*

10.1 Introduction and motivation

Generally, the mechanism of hypersurface evolution in a differential process is investigated through changes in hypersurface profiles with process time. The nature of hypersurface changes and how the hypersurface evolves over time are not yet entirely known in the geometric sense.

Hypersurface processing tools and techniques are widespread in computer graphics, animation, medical imaging, computer aided modelling, and computer vision [8]. In recent years, there has been an increasing interest in hypersurface deformation (motion) governed by geometric PDEs. This Chapter introduces a new approach to modelling geometrical flows of contours and hypersurfaces having in view future topics in optimization problems.

Differential evolution has shown to be very efficient when solving global optimization problems with simple bounds. In this Chapter, we propose a modified constrained differential evolution based on different Tzitzeica constraints handling techniques, namely, feasibility and dominance rules, especially for geometric programming and reliability optimal allocation [13].

In Sections $2-8$ we are interested both in the original forms and in the ending forms, beneficial to some optimization problems.

The first centro-affine invariant was introduced in 1907 by George Tzitzeica in the classical theory of surfaces. Namely, if M is a (non-flat) surface embedded in $\mathbb{R}^3$, the Tzitzeica invariant is $Tz(M) = \frac{K}{d^4}$, where K is the Gaussian curvature and d is the distance from the origin of $\mathbb{R}^3$ to the tangent plane in an arbitrary point of M [9], [10]. Since then, the class of surfaces M with a constant $Tz(M)$ was the subject of several papers, using the names *Tzitzeica surfaces* (especially by Romanian geometers), *affine spheres* (by Blaschke and his followers) and *projective spheres* (by Wilczynski and his followers).

Generalization: a hypersurface M of $\mathbb{R}^{n+1}$ is called Tzitzeica hypersurface if $Tz(M) = \frac{K}{d^{n+2}}$ is a constant.

10.2 Convexity of Tzitzeica hypersurfaces

The Tzitzeica surfaces theory is well-known [1], [3], [6], [8] - [10].

The constant level sets $\prod_{i=1}^{n} x_i = c$ attached to the function

$$f\colon \mathbb{R}^n \to \mathbb{R},\ f(x) = \prod_{i=1}^{n} x_i,\ x = (x_1, ..., x_n),$$

are Tzitzeica hypersurfaces in $\mathbb{R}^n$. Each of these hypersurfaces reduces to: union of 2^{n-1} disjoint and closed sets if $c \neq 0$, union of coordinate hyperplanes if $c = 0$.

Proposition 10.2.1. *Each connected component of the Tzitzeica hypersurface $\prod_{i=1}^{n} x_i = c$ is convex ([12]).*

Proof. Since

$$\frac{\partial^2 f}{\partial x_i^2} = 0,\ \frac{\partial^2 f}{\partial x_i \partial x_j} = \frac{f}{x_i x_j},$$

we find

$$d^2 f = 2f \sum_{i<j} \frac{dx_i}{x_i} \frac{dx_j}{x_j} = f\left(\left(\sum_{i=1}^{n} \frac{dx_i}{x_i}\right)^2 - \sum_{i=1}^{n} \frac{dx_i^2}{x_i^2}\right).$$

Suppose $c \neq 0$. The restriction of this Hessian to the tangent plane $\sum_{i=1}^{n} \frac{dx_i}{x_i} = 0$, i.e.,

$$d^2 f = -c\left(\sum_{i=1}^{n} \frac{dx_i^2}{x_i^2}\right)$$

is negative or positive definite according to the sign of c. Consequently, the second fundamental form is semi-definite. □

10.3 Normal evolution of Tzitzeica hypersurfaces

The monographs [2], [5], [8] offer the reader a treatment of the theory of evolution PDEs. The monographs [4], [6], [7] contain related topics. All of these suggested the following ways of deformation.

In relation to deformation of Tzitzeica surfaces, we have to say from the beginning what is well-known [9], [10]: *the Tzitzeica property is preserved only by affine transformations and centro-affine ones.*

10.3.1 Steepest increase lines

Let $f(x) = \prod_{i=1}^{n} x_i$ be the Tzitzeica function. The *gradient vector field* $\nabla f = \left(\dfrac{f}{x_i}\right)$ has two properties: (i) shows at every point $(x_1, ..., x_n)$ the direction and the sense of steepest increase of f; (ii) is orthogonal to the constant level sets of f (Tzitzeica hypersurfaces). Consequently, the field lines of ∇f, i.e., the solutions of the ODE system $\frac{dx_i}{dt} = \frac{c}{x_i}$, $i = 1, ..., n$ are very important. Since $x_1 dx_1 = ... = x_n dx_n = c\, dt$, the steepest increase orbits are intersections of hyperbolic hypercylinders $x_2^2 - x_1^2 = c_2, ..., x_n^2 - x_1^2 = c_n$ (see first integrals).

The field ∇f is solenoidal, and consequently its flow conserves the volume.

10.3.2 Evolution along the normal vector field

Let us consider the Tzitzeica hypersurface $f(x) = x_1 \cdots x_n = c$ and the vector field $N = \nabla f$ with the components $\left(\frac{c}{x_1}, ..., \frac{c}{x_n}\right)$. On the given Tzitzeica hypersurface, the field N is a normal vector field.

Theorem 10.3.1. *A normal deformation of a Tzitzeica hypersurface* $x_1 \cdots x_n = c$ *can be written as*

$$(x_1^2 - 2ct)(x_2^2 - 2ct) \cdots (x_n^2 - 2ct) = c^2.$$

Proof. The evolution of the surface $f(x) = c$ along the direction of the normal vector field N, i.e., with respect to a parameter $t \in [0, \epsilon)$, is made taking into consideration the extension (variation) function $\mathcal{F}(x_1, ..., x_n, t)$ as a solution of the Cauchy problem

$$\frac{\partial \mathcal{F}}{\partial t} = -\langle N, \nabla \mathcal{F}\rangle, \ \mathcal{F}(x_1, ..., x_n, 0) = f(x_1, ..., x_n) - c.$$

Explicitly the PDE is

$$\frac{\partial \mathcal{F}}{\partial t} + \frac{c}{x_1}\frac{\partial \mathcal{F}}{\partial x_1} + ... + \frac{c}{x_n}\frac{\partial \mathcal{F}}{\partial x_n} = 0,$$

with the characteristic system

$$\frac{dt}{1} = x_1 \frac{dx_1}{c} = ... = x_n \frac{dx_n}{c}.$$

We use the first integrals (general solution of the characteristic system)

$$x_1^2 - 2c\,t = c_1, x_2^2 - x_1^2 = c_2, ..., x_n^2 - x_1^2 = c_n.$$

Let $\Phi : \mathbb{R}^n \to \mathbb{R}$ be an arbitrary C^1 function. Then

$$\mathcal{F}(x_1, ..., x_n, t) = \Phi(x_1^2 - 2ct, x_2^2 - x_1^2, ..., x_n^2 - x_1^2),$$

together with the initial data

$$\mathcal{F}(x_1, ..., x_n, 0) = f(x_1, ..., x_n) = x_1 \cdots x_n - c.$$

The condition

$$\Phi(x_1^2, x_2^2 - x_1^2, ..., x_n^2 - x_1^2) = x_1 \cdots x_n - c$$

implies

$$\Phi(y_1, y_2, ..., y_n) = \sqrt{y_1(y_2 + y_1) \cdots (y_n + y_1)} - c.$$

We find the solution

$$\mathcal{F}(x_1, ..., x_n, t) = \sqrt{(x_1^2 - 2ct)(x_2^2 - 2ct) \cdots (x_n^2 - 2ct)} - c$$

and the result follows. □

The normal deformation of a Tzitzeica hypersurface is not a Tzitzeica hypersurface since the deformation is realised by the diffeomorphism $y_i = x_i^2 - 2ct$, $i = 1, ..., n$, which is not a centro-affine one.

Remark 10.3.1. *(i) The evolution Hamilton-Jacobi PDE*

$$\frac{\partial \mathcal{F}}{\partial t} = \pm\langle \nabla_x \mathcal{F}, \nabla_x \mathcal{F} \rangle$$

describes the movement of the surface $\mathcal{F}(x_1, ..., x_n; t) = 0$ along its normal vector field $\pm\nabla_x \mathcal{F}$.

(ii) Let $F(x, t)$ be a C^1 function. The ODE $\frac{dx}{dt}(t) = \nabla_x F(x(t), t)$, gives the PDE

$$\frac{d}{dt} F(x(t), t) = ||\nabla_x F||^2 + \frac{\partial F}{\partial t}.$$

Consequently, the hypersurfaces in the flow evolution (i) are first integrals of ODE (ii).

10.4 Infinitesimal normal transformation of Tzitzeica hypersurface

Our aim is to deform a Tzitzeica hypersurface $x_1 \cdots x_n = c$ by infinitesimal transformation associated to its normal vector field $N = \left(\frac{c}{x_1}, ..., \frac{c}{x_n}\right)$. Between the deformed hypersurfaces does there exist one which is Tzitzeica?

Theorem 10.4.1. *The infinitesimal transformation of the normal flow deforms the Tzitzeica hypersurface $x_1 \cdots x_n = c$ into the hypersurface*

$$\prod_{i=1}^{n} \frac{y_i \pm \sqrt{y_i^2 - 4ct}}{2} = C.$$

Proof. The infinitesimal transformation produced by the normal vector field N is

$$y_1 = x_1 + t\,\frac{c}{x_1}, ..., y_n = x_n + t\,\frac{c}{x_n}, \quad x_1 \cdots x_n = c, \;\; t \in [0, \varepsilon).$$

One determines $x_1, ..., x_n$ (one reverses the infinitesimal transformation),

$$x_i = \frac{y_i \pm \sqrt{y_i^2 - 4ct}}{2}, \; i = 1, ..., n$$

(positive roots) and we introduce in the first equation. □

These new hypersurfaces are obtained from a Tzitzeica hypersurface using a diffeomorphism different from a centro-affine one. Consequently, they are not Tzitzeica hypersurfaces.

10.5 Evolution along a centro-affine vector field

Let us consider the Tzitzeica hypersurface $f(x) = x_1 \cdots x_n = c$ and the centro-affine vector field Ax, det $A \neq 0$. The evolution along this vector field is described by the flow Cauchy problem

$$\frac{\partial \mathcal{F}}{\partial t} = -\langle Ax, \nabla \mathcal{F} \rangle, \; \mathcal{F}(x_1, ..., x_n, 0) = f(x_1, ..., x_n) - c.$$

The characteristic system

$$\frac{dt}{1} = \frac{dx_1}{A_1 x} = ... = \frac{dx_n}{A_n x},$$

i.e., $\frac{dx}{dt} = Ax$, has the general solution $x = e^{At}C$. Then

$$\mathcal{F}(x_1, ..., x_n, t) = \Phi(e^{-At}x),$$

$$\mathcal{F}(x_1, ..., x_n, 0) = f(x_1, ..., x_n) - c = x_1 \cdots x_n - c.$$

It follows the implicit equation of deformed hypersurface

$$\mathcal{F}(x_1, ..., x_n, t) = (e^{-At})_1\, x \cdots (e^{-At})_n\, x - c = 0.$$

The centro-affine diffeomorphism $y = e^{-At}\, x$ shows that our deformation gives a Tzitzeica hypersurface.

Problem Let $B(t), B(0) = I$ be a given family of centro-affine diffeomorphisms. There does exists one centro-affine vector field Ax that creates the given family $B(t)$?

Proposition 10.5.1. *A given family of centro-affine dieomorphisms $B(t), B(0) = I$ is one of exponential type if and only if it is solution of matrix ODE : $\dot{B}(t)^2 - B(t)\ddot{B}(t) = 0$.*

Proof. By hypothesis, $B(t)$ is invertible and has positive determinant. Suppose that there exists a constant matrix A such that $e^{-At} = B(t)$. We find $A = -B^{-1}(t)\dot{B}(t)$. Since the left-hand member is independent on t, we obtain, by derivation with respect to t, the matrix ODE: $B^{-2}\dot{B}(t)^2 - B^{-1}(t)\ddot{B}(t) = 0$, which is equivalent to the one in the statement. Conversely, the matrix equation $\dot{B}(t)^2 - B(t)\ddot{B}(t) = 0$ has the solution $B(t) = e^{Ct}$. It follows that $A = -C$. Consequently, the Proposition is true. □

10.6 Evolution along an affine vector field

Let us consider the Tzitzeica hypersurface $f(x) = x_1 \cdots x_n = c$ and the affine vector field $Ax + h$, $\det A \neq 0$. The evolution along this vector field is described by the flow Cauchy problem

$$\frac{\partial \mathcal{F}}{\partial t} = -\langle Ax + h, \nabla \mathcal{F} \rangle,\ \mathcal{F}(x_1, ..., x_n, 0) = f(x_1, ..., x_n) - c.$$

The associated characteristic system

$$\frac{dt}{1} = \frac{dx_1}{A_1x + h_1} = ... = \frac{dx_n}{A_nx + h_n},$$

i.e., $\frac{dx}{dt} = Ax + h$, has the general solution

$$x = e^{At}\left\{C + \left(I - e^{-At}\right)h\right\}.$$

Then

$$\mathcal{F}(x_1, ..., x_n, t) = \Phi\left(e^{-At}x - \left(I - e^{-At}\right)h\right),$$

$$\mathcal{F}(x_1, ..., x_n, 0) = f(x_1, ..., x_n) - c = x_1 \cdots x_n - c.$$

It follows

Theorem 10.6.1. *The implicit equation of deformed hypersurface is*

$$\mathcal{F}(x_1, ..., x_n, t) = \prod_{i=1}^{n}[(e^{-At}x)_i - ((I - e^{-At})h)_i] - c = 0.$$

The affine diffeomorphism $y = e^{-At}x - (I - e^{-At})h$ shows that our deformation is a Tzitzeica hypersurface, with respect to a new center point $(e^{-At} - I)h$.

Let us highlight an affine diffeomorphism that preserves the quality of Tzitzeica hypersurface, like in duality between series and parallel systems in reliability [1]. Let $x_1 \cdots x_n = c$ be a Tzitzeica hypersurface, with the center $(0, ..., 0)$. The affine diffeomorphism (symmetry and translation) $x_1 = 1 - y_1, ..., x_n = 1 - y_n$ changes the given Tzitzeica hypersurface into a Tzitzeica hypersurface $(1-y_1) \cdots (1-y_n) = c$, with the center $(1, ..., 1)$. Here, the associated matrix has the determinant ± 1 (equi-affine diffeomorphism). In our context, a well-known result [9], [10] can be written in the general form

Theorem 10.6.2. *Let $x = (x_1, ..., x_n)$ and $y = (y_1, ..., y_n)$ be generic vectors in $\mathbb{R}^n$. An affine diffeomorphism $x = Ay + b$ changes the Tzitzeica hypersurface $x_1 \cdots x_n = c$, with the center $(0, ..., 0)$, into a Tzitzeica hypersurface $(Ay_1 + b_1) \cdots (Ay_n + b_n) = c$, with the center $A^{-1}b$.*

10.7 Tzitzeica hypersurfaces as invariants with respect to excess demand flow

We extend the theory of single-time dynamics produced by the excess demand vector field (see [3], [7]), from 3 to $n=$odd dimensions.

Let $p = (p_1, ..., p_n)$ be the vector of prices. The general *excess demand vector field* E has the components

$$\begin{array}{lllllllll} E_1(p) & = & -\frac{p_2}{p_1+p_2} & + & \frac{p_3}{p_1+p_3} & - & ... & + & (-1)^{n-1}\frac{p_n}{p_1+p_n} \\ E_2(p) & = & -\frac{p_3}{p_2+p_3} & + & \frac{p_4}{p_2+p_4} & - & ... & + & (-1)^{n-1}\frac{p_1}{p_2+p_1} \\ ... & ... & ... & & ... & & ... & & ... \\ E_n(p) & = & -\frac{p_1}{p_n+p_1} & + & \frac{p_2}{p_n+p_2} & - & ... & + & (-1)^{n-1}\frac{p_{n-1}}{p_n+p_{n-1}}. \end{array}$$

Using these data, we can write the *excess demand flow* $\frac{dp}{dt}(t) = E(p(t))$.

The positive homogeneity of degree zero of each component $E_i(p)$ means that if all prices are multiplied by the same positive constant, the excess demands do not vary (is invariant with respect to homothetic transformations).

The *Walras Law* $\sum_{i=1}^{n} p_i E_i = 0$ means that

$$f_1(p) = \frac{1}{2}\left(p_1^2 + p_2^2 + ... + p_n^2\right)$$

is a first integral. The associated constant level sets of $f_1(p)$ are spheres (compact hypersurfaces, invariant under the excess demand flow). Consequently, the vector field E is complete.

Let us introduce and explain what is a *Tzitzeica Law.*

Theorem 10.7.1. *The Tzitzeica function $f(p) = p_1 \cdots p_n$ is a first integral for the excess demand flow.*

Proof. We can verify the relation

$$\sum_{CP} p_1 p_2 \cdots p_{n-1} E_n = 0 \text{ (CP = summation by cyclic permutation).}$$

Indeed, for n = odd number, we can write

$$E_1(p) = p_1\left(\frac{1}{p_1+p_2} - \frac{1}{p_1+p_3} + ...\right)$$

$$E_2(p) = p_2\left(\frac{1}{p_2+p_3} - \frac{1}{p_2+p_4} + ...\right)$$

....

$$E_n(p) = p_n \left(\frac{1}{p_n + p_1} - \frac{1}{p_n + p_2} + ... \right),$$

whence the previous relation follows. Consequently, a second first integral of the flow is $f_2(p) = p_1 \cdots p_n$, i.e., the excess demand flow conserves the standard Tzitzeica hypersurfaces. □

Practically this means that the "volume of the prices" is constant.

Remark 10.7.1. *For $n =$ odd number, we can write*

$$\begin{pmatrix} E_1(p) \\ ... \\ E_n(p) \end{pmatrix} = A(p) \begin{pmatrix} p_1 \\ ... \\ p_n \end{pmatrix},$$

where A is a skew-symmetric $n \times n$ matrix.

The previous flow and the Euclidean metric determine the least squares Lagrangian

$$L = \frac{1}{2}\delta^{ij} \left(\frac{dp_i}{dt} - E_i \right) \left(\frac{dp_j}{dt} - E_j \right).$$

Its Euler-Lagrange ODEs are

$$\frac{d^2 p_i}{dt^2} = \delta^{jk} \left(\frac{\partial E_i}{\partial p_j} - \frac{\partial E_j}{\partial p_i} \right) \frac{dp_k}{dt} + \frac{\partial f}{\partial p_i},$$

where $f = \frac{1}{2}\delta^{ij} E_i E_j$ is the *energy density*. The last ODEs describe the generated geometric dynamics (geodesic motion in a gyroscopic field of forces) that includes the excess demand flow.

10.8 General evolution of Tzitzeica surfaces

A parametric Tzitzeica surface in $\mathbb{R}^3$ is characterized by the completely integrable PDE system

$$r_{uu} = \frac{h_u}{h}\, r_u + \frac{1}{h}\, r_v, r_{uv} = h\, r,\ r_{vv} = \frac{1}{h}\, r_u + \frac{h_v}{h}\, r_v,\ (\ln h)_{uv} = h - \frac{1}{h^2}.$$

Theorem 10.8.1. *Let $r(u,v)$ be a Tzitzeica surface. An associated evolution surface $r(u,v,t)$, $t \in [0,\epsilon)$ is a Tzitzeica surface if and only if $\frac{\partial r}{\partial t}(u,v,t) = A(t)r(u,v,t)$, where $A(t)$ is a non-degenerate matrix.*

Proof. Let $(u,v) \to r(u,v,t)$, $t \in [0,\epsilon)$ be a variation of the Tzitzeica surface $r(u,v) = r(u,v,0)$.

If we consider $\frac{\partial r}{\partial t}(u,v,t) = A(t)r(u,v,t)$, with $A(t)$ a non-singular matrix, then using the fundamental matrix $\Omega(t)$, we find the solution $r(u,v,t) = \Omega(t)r(u,v,0) = \Omega(t)r(u,v)$, which will satisfy the Tzitzeica system, and hence the evolution is a Tzitzeica surface.

Conversely, let us take the evolution surface $r(u,v,t)$ as Tzitzeica for any $t \in [0,\epsilon)$. By partial derivative with respect to t, we find

$$\frac{\partial}{\partial t}r_{uu} = \frac{h_u}{h}\frac{\partial}{\partial t}r_u + \frac{1}{h}\frac{\partial}{\partial t}r_v, \quad \frac{\partial}{\partial t}r_{uv} = h\frac{\partial}{\partial t}r, \quad \frac{\partial}{\partial t}r_{vv} = \frac{1}{h}\frac{\partial}{\partial t}r_u + \frac{h_v}{h}\frac{\partial}{\partial t}r_v.$$

Hence the surface $\frac{\partial}{\partial t}r(u,v,t)$ is again a Tzitzeica surface. According to the well-known theory, there exists a time dependent centro-affine transformation, which applies the surfaces one to another, i.e., $\frac{\partial r}{\partial t}(u,v,t) = A(t)r(u,v,t)$. □

Remark 10.8.1. *If the system of PDEs is linear in $r(u,v)$, then an evolution $\frac{\partial r}{\partial t}(u,v,t) = A(t)r(u,v,t)$ produces surfaces of the same type.*

10.9 Self-evaluation problems

Problem 99. *Show that there exists a constant level set of a solution of PDE $yzF_x + zxF_y + xyF_z = 3xyz$ that is Tzitzeica surface.*

Hint A particular solution, $2F(x,y,z) = x^2 + y^2 + z^2$.

Problem 100. *Can the graph of a solution of the PDE $2u_x + 3u_y + 8u = 0$ be a Tzitzeica surface?*

Hint The general solution of the PDE is $u(x,y) = e^{-4x}g(3x-2y)$, where g is a differentiable function defined over the real line.

Problem 101. *Study different types of evolutions of the following plane curves: (1) Devil curve $y^2(y^2-a^2) = x^2(x^2-b^2)$; (2) Folium of Descartes $x^3+y^3-3axy = 0,\ a > 0$; Serpentine curve $x^2y+a^2y-abx = 0,\ ab > 0$.*

Problem 102. *Study different types of evolutions of the following surfaces: (1) Fermat cubic* $x^3 + y^3 + z^3 = 1$*; (2) monkey saddle* $z = x^3 - 3xy^2$*; (3) Whitney umbrella* $x^2 - y^2 z = 0$.

Problem 103. *Study different types of evolutions of the solutions to the following ODEs: (1) Blasius ODE* $\frac{d^3y}{dx^3} + y\frac{d^2y}{dx^2} = 0$*; (2) Langmuir - Blodgett ODE* $\sqrt{y}\,\frac{d^2y}{dx^2} = e^x$*; (3) Painleve ODE* $\frac{d^2y}{dx^2} = 6y^2 + t$*; (4) Thomas - Fermi ODE* $\frac{d^2y}{dx^2} = \frac{1}{\sqrt{x}}\, y^{3/2}$.

Problem 104. *Study different types of evolutions of the solutions to the following PDEs: (1) Bateman - Burgers PDE* $u_t + uu_x = au_{xx}$*; (2) Dym PDE* $u_t = u^3 u_{xxx}$*; (3) Fisher PDE* $u_t = u(1-u) + u_{xx}$*; (4) Sinh - Gordon PDE* $u_{xt} = \sinh u$*; (5) Heisenberg ferromagnet PDEs* $S_t = S \times S_{xx}$ *on* $\mathbb{R}^3$*; (6) Minimal surface PDE* $\partial_x \frac{u_x}{\sqrt{1+u_x^2+u_y^2}} + \partial_y \frac{u_y}{\sqrt{1+u_x^2+u_y^2}} = 0$ *(given* $u(x,y)$ *on* $\mathbb{R}^2$*, show that there are numbers* a, b, c *such that* $u(x,y) = ax + by + c$*).*

Problem 105. *Study different types of evolutions of the Tzitzeica revolution hypersurface*

$$M : x_{n+1}\left[x_1^2 + ... + x_n^2\right]^{\frac{n}{2}} = 1.$$

Hint $Tz(M) = \frac{(-n)^n}{(n+1)^{n+1}}$.

Bibliography

[1] S. A. Abed, C. Udriste, I. Tevy, (2017). *Optimal reliability allocation for redundancy series-parallel systems*, Eur. J. Pure Appl. Math., 10(4), 873-885.

[2] A. F. Agnew, A. Bobe, W. G. Boskoff, B. D. Suceava, (2010). *Tzitzeica curves and surfaces.* The Mathematica Journal, Wolfram Media, Inc. 12.

[3] S. Antontsev, S. Shmarev, (2015). *Evolution PDEs with Nonstandard Growth Conditions*, Springer.

[4] M. T. Calapso, C. Udriste, (2005). *Tzitzeica and Walras laws in geometric economics.* Analele Universităţii Bucureşti 55(1), 213-218.

[5] O. Calin, C. Udriste, (2014). *Geometric Modeling in Probability and Statistics*, Springer.

[6] O. Constantinescu, M. Crasmareanu, (2011). *A new Tzitzeica hypersurface and cubic Finslerian metrics of Berwald type*, Balkan J. Geom. Appl., 16, 2 (2011), pp. 27-34.

[7] L. C. Evans, (1998). *Partial Differential Equations*, Graduate Texts in Mathematics 19, AMS.

[8] S. Morig, (2010). *Geometric surface evolution with tangential contribution*, Journal of Computational and Applied Mathematics, 233, 5, 1277-1287.

[9] C. Udriste, (1994). *Convex Functions and Optimization Methods on Riemannian Manifolds*, Kluwer Academic Publisheres.

[10] C. Udriste, M. Ferrara, D. Opris, (2004). *Economic Geometric Dynamics*, Geometry Balkan Press.

[11] C. Udriste, (2005). *Tzitzeica theory - opportunity for reflection in Mathematics.* Balkan J. Geom. Appl. 10(1), 110-120.

[12] C. Udriste, S. A. Abed, A. S. Rasheed, (2016). *Optimal reliability allocation.* American Review of Mathematics and Statistics 10, 82-91. DOI:10.15640/arms.

[13] C. Udriste, S. A. Abed, I. Tevy, (2017). *Geometric programming approaches of reliability allocation.* U.P.B. Sci. Bull., Series A, 79 (3), 3-10.

[14] C. Udriste, I. Tevy, A. S. Rasheed, (2018). *Evolution of Tzitzeica hypersurfaces*, Atti Accad. Pelorit. Pericol. Cl. Sci. Fis. Mat. Nat., 96, 1, A7.

[15] G. Tzitzeica, (1907). *Sur une nouvelle classe de surfaces.* Les Comptes Rendus de l'Académie des Sciences 144, 1257-1259.

[16] G. Tzitzeica, (1908); (1909). *Sur une nouvelle classe de surfaces.* Rendiconti del Circulo Matematico di Palermo 25, 180-187; 28, 210-216.

Chapter 11

EVOLUTIVE OPTIMIZATION PROBLEMS

Motto:
"I was looking upwards randomly -
In a wild splendour
I saw the mountain Ceahlau west,
Far away, in the blue sky,
A sunny giant,
Guard to our country."

George Coşbuc - *The Summer*

This is a Chapter about continuous time deterministic evolutionary dynamics for optimization problems. The topics include: (i) the evolution of the objective function; (ii) the evolution of the constraints; (iii) the evolution of the minimum value; (iv) the evolution of the catastrophe manifold; (v) the evolution of an optimal control problem. For the context of optimal control problems, we also described: (1) a general theory about evolution of a curve described by an ODE; (2) a general theory about evolution of a surface described by a PDE; (3) evolution of a plane curve following a first order PDE. Suggestions for further reading: [1]-[9].

11.1 Evolution of an optimization problem

In general terms, "optimization problems that change over time" are called dynamic problems/time - dependent problems. The first idea about the evolution of a free optimization problem appeared in [5]. Now this subject can be included in the geometric evolution of ODEs and PDEs [1],[3], [4]. Related topics are found in [2], [6], [7].

In the *trade-off analysis of an optimum problem* we vary the constraints, and see the effect on the optimal value of the problem. The *sensitivity analysis of an optimum problem* is closely related to the trade-off analysis. In sensitivity analysis, we consider how small changes in the constraints affect the optimal objective value. Both problems reflect the idea that in many practical problems, the constraints are not really set in stone, and can be changed, especially if there is a compelling reason to do so (such as a drastic improvement in the obtained objective). We extend the trade-off analysis and the sensitivity analysis to evolution of an optimization problem.

Let $x = (x^1, ..., x^n) \in \mathbb{R}^n$ be the vector of decision variables, $f\colon \mathbb{R}^n \to \mathbb{R}$ be a C^2 objective function and $g_i\colon \mathbb{R}^n \to \mathbb{R}$, $i = 1, ..., k$, be C^2 functions that describe the constraints. A general optimization problem is of the form:

$$\min_x f(x)$$

subject to

$$g_1(x) \leq 0, ..., g_k(x) \leq 0.$$

The t-evolution of the graph of f and the t-evolution of the set described by constraints have a geometrical sense and are very clear. Here we solve the following problem: what is the significance of the t-evolution of an optimization problem? Of course, though the transition from the evolution single-time parameter t to the evolution multi-time parameter $t = (t^1, ..., t^m)$ is not so difficult, the multi-time case has its own specifical problems and is still in research.

11.1.1 Evolution of objective function

Without loss of generality, we consider a simplified problem of optimization:

$$\min_x f(x) \text{ subject to } g(x) = 0.$$

Suppose the objective function $f\colon \mathbb{R}^n \to \mathbb{R}$ in this problem is smooth and regular. Its graph $G(f)\colon y = f(x)$ is a submanifold in $\mathbb{R}^{n+1}$, geometrically characterized by: the metric g_{ij}, or its inverse g^{ij},

$$g_{ij} = \delta_{ij} + f_i f_j,\ g^{ij} = \delta^{ij} - \frac{f_i f_j}{1 + \|\nabla f\|^2},$$

the normal versor field n, the second fundamental form h_{ij},

$$n = \frac{((f_i), -1)}{\sqrt{1 + \|\nabla f\|^2}},\ h_{ij} = \frac{f_{ij}}{\sqrt{1 + \|\nabla f\|^2}},$$

the mean curvature H, and the Gauss curvature K,

$$H = div\left(\frac{\nabla f}{\sqrt{1 + \|\nabla f\|^2}}\right),\ K = \det \frac{\nabla^2 f}{(1 + \|\nabla f\|^2)^{\frac{n+2}{2}}}.$$

The graph submanifold is described by the embedding vector field $X(x) = (x, f(x))$. Therefore, its evolution in time, along the normal vector, is introduced by a variation

$$F = F(x,t), F(x,0) = f(x),\ t \in [0,T]$$

and the associated vector field $Y(x,t) = (x, t, F(x,t))$, which satisfies the evolution PDE

$$\partial_t Y(x,t) = -H(x,t) n(x,t),\ Y(x,0) = X(x),$$

where $H(x,t)$ is the mean curvature. The last PDE is equivalent to a flow PDE. It follows

Proposition 11.1.1. *The normal evolution F of objective function f is described by the flow Cauchy problem*

$$\dot{F} = \sqrt{1 + \|\nabla F\|^2}\ H,\ F(x,0) = f(x).$$

11.1.2 Evolution of a constraint manifold

Let $S : g(x) = 0$ be a regular implicit hypersurface representing the constraint in an optimization problem

$$\min_x f(x) \text{ subject to } g(x) = 0.$$

Suppose $x = x(u)$, $u \in \mathbb{R}^{n-1}$, is a parametrization of S, i.e., $g(x(u)) = 0$, $\forall u \in \mathbb{R}^{n-1}$. The time evolution hypersurface $S(t)$ is described by the implicit equation

$$S(t)\colon G(x(u,t),t) = 0,\ u \in \mathbb{R}^{n-1},\ t \in [0,T],\ S(0) = S,$$

i.e., $G(x,0) = g(x)$. Using $X = x(u,t)$, and derivation with respect to t, we find that the function G must verify the relation

$$\langle \nabla G, X_t \rangle + G_t = 0$$

or, explicitly, the PDE

$$\frac{\partial G}{\partial x^i}\frac{\partial x^i}{\partial t} + \frac{\partial G}{\partial t} = 0,$$

with the unknown function G. If we impose $\frac{\partial}{\partial t}x = \alpha Y$, then we find the PDE $G_t = -\alpha\langle \nabla G, Y\rangle$, with unknown G, fixed by the initial condition.

Let $n = \frac{\nabla G}{||\nabla G||}$ be the normal versor to the hypersurface $S(t)$. If we accept the evolution along the normal versor field, i.e., $\frac{\partial}{\partial t}x = \beta\, n$, then we get $G_t = -\beta||\nabla G||$. For scaling, some authors introduce $\beta = K$ (Gauss curvature) or $\beta = H$ (mean curvature), attending interesting nonlinear second order PDEs in the unknown G.

Summing up, we find

Proposition 11.1.2. *The normal evolution G of constraint function g is described by the flow Cauchy problem $G_t = -\beta||\nabla G||$, $G(x,0) = g(x)$.*

Giving G, let us look at the relationship

$$\langle \nabla G, X_t \rangle + G_t = 0$$

as an equation with the unknown X_t. Finding a particular solution for the normal component, we obtain the normal velocity

$$X_t = -G_t(x,t)\ \frac{\nabla G(x,t)}{||\nabla G(x,t)||^2}.$$

The general solution is

$$X_t = -G_t(x,t)\ \frac{\nabla G(x,t)}{||\nabla G(x,t)||^2} + \lambda(x,t)V(x,t),$$

where $\langle \nabla G(x,t), V(x,t)\rangle = 0$, i.e., the vector field $V(x,t)$ is tangent to the hypersurface $S(t)$, and $\lambda(x,t)$ is an arbitrary scalar field.

11.1.3 Evolution of minimum value

Suppose that x^* and $f^* = f(x^*)$ is a solution of the problem

$$\min_x f(x) \text{ subject to } g(x) = 0.$$

Let us analyse what happens with these data during the evolution.

By t-evolution, the Lagrange function $l(x) = f(x) + \lambda g(x)$ becomes

$$L(x, t, \lambda) = F(x, t) + \lambda(t) G(x, t).$$

Theorem 11.1.1. *The minimum value F^* satisfies the relation*

$$\frac{dF^*}{dt} - \frac{\partial F^*}{\partial t} = \lambda(t)\ \frac{\partial G^*}{\partial t}.$$

Proof. We use the Lagrange function

$$L(x, t, \lambda) = F(x, t) + \lambda G(x, t).$$

The critical point condition $\frac{\partial L}{\partial x} = 0$ gives $x = x(\lambda, t)$. From the constraint condition $G(x, t) = 0$, i.e., $G(x(\lambda, t), t) = 0$, we find $\lambda = \lambda(t)$ and finally $x(t) = x(\lambda(t), t)$. Consequently, the minimum value satisfies the relation in Theorem. □

Remark 11.1.1. *For example, if $F(x, t) = f(x)$ and $G(x, t) = g(x) - t$, then the previous PDE reduces to the well-known relation*

$$\frac{df^*}{dt} = -\lambda(t).$$

In other words, a classical optimization problem "$f(x)$=extremum, subject to $g(x) = c$" is a particular case of evolution of the optimization problem "$f(x)$=extremum, subject to $g(x) = 0$".

If $F(x, t)$ is combined to $G(x, t) = g(x)$, then we get

$$\frac{df^*}{dt} = \frac{\partial f^*}{\partial t}$$

(see also the properties of "minimum functions" [10]).

If G is a vectorial function, then in the relation of Theorem, the right hand term becomes a scalar product. The optimization problem "$f(x)$=extremum, subject to $g_1(x) = t,\ g_2(x) = t$" leads to

$$\frac{df^*}{dt} = -(\lambda_1(t) + \lambda_2(t)).$$

11.2 Evolution of the catastrophe manifold associated to a Lagrange potential

We start with a Lagrange function

$$L : \mathbb{R}^n \times \mathbb{R}^k \to \mathbb{R},\ L(x,\lambda) = f(x) + \langle \lambda, g(x) \rangle,$$

where f and g are of class C^2. The point $x \in \mathbb{R}^n$ is called *state.* The point $\lambda \in \mathbb{R}^k$ is called *control.* The partial function $x \to L(x,\lambda)$ is called *potential.*

The set of all critical points $M \subset \mathbb{R}^n \times \mathbb{R}^k$ of the potentials $x \to L(x,\lambda)$, characterized by the algebraic system

$$\frac{\partial f}{\partial x^i} + \langle \lambda, \frac{\partial g}{\partial x^i} \rangle \ = 0,\ i = 1, ..., n,$$

is called the *catastrophe manifold.*

Let $\pi : \mathbb{R}^n \times \mathbb{R}^k \to \mathbb{R}^k$, $\pi(x,\lambda) = \lambda$ be the natural projection. The restriction of π to M is denoted by χ and is called *catastrophe application.* The subset $S \subset M$ consisting of the singular points, with respect to x, of the catastrophe application $\chi : M \to \mathbb{R}^k$, i.e., the set of points at which the rank of the Jacobian matrix $J_x(\chi)$ is smaller than k, is called the *singularity set.* The singularity set S is described by $n+1$ equations

$$\frac{\partial f}{\partial x^i} + \langle \lambda, \frac{\partial g}{\partial x^i} \rangle \ = 0,\ \det \left(\frac{\partial^2 L}{\partial x^i \partial x^j} \right) = 0.$$

The image $B = \chi(S) \subset \mathbb{R}^k$ is called *bifurcation set.* The bifurcation set B is the set where the number and nature of critical points change; the change takes place only when passing through a degenerate critical point.

11.2.1 Evolution preserving the catastrophe manifold

Now we introduce a variation parameter $\epsilon \in [0,T)$ independent of x. Then it appears the perturbed functions $f(x,\epsilon)$, $\lambda(\epsilon)$, $g(x,\epsilon)$. We impose the initial conditions

$$f(x,0) = f(x),\ \lambda(0) = \lambda,\ g(x,0) = g(x).$$

Also, we need a new Lagrange function

$$L(x, \lambda(\epsilon), \epsilon) = f(x; \epsilon) + \langle \lambda(\epsilon), g(x; \epsilon) \rangle$$

and the associated critical point condition

$$\left(\frac{\partial f}{\partial x^i} + \langle \lambda, \frac{\partial g}{\partial x^i} \rangle \right) (x, \lambda(\epsilon), \epsilon) \ = 0.$$

To continue, we must declare the variational objects

$$\frac{\partial f}{\partial \epsilon}|_{\epsilon=0} = \xi(x), \ \frac{\partial g}{\partial \epsilon}|_{\epsilon=0} = \eta(x), \ \frac{\partial \lambda}{\partial \epsilon}|_{\epsilon=0} = \varsigma.$$

Theorem 11.2.1. *Let $x(\lambda)$ be a critical point. The vector $\nabla\xi(x(\lambda))$ belongs to the space generated by $\nabla\eta(x(\lambda))$ and $\nabla g(x(\lambda))$, i.e.,*

$$\frac{\partial \xi}{\partial x^i} + \langle \lambda, \frac{\partial \eta}{\partial x^i} \rangle + \langle \varsigma, \frac{\partial g}{\partial x^i} \rangle \ = 0.$$

Proof. **Variant 1** We start from the critical point condition. Taking the derivative with respect to ϵ, and setting $\epsilon = 0$, we find the equations on variation in the Theorem.

Variant 2 We set

$$f(x, \epsilon) = f(x) + \epsilon\xi(x) + o(\epsilon),$$
$$g(x, \epsilon) = g(x) + \epsilon\eta(x) + o(\epsilon),$$
$$\lambda(\epsilon) = \lambda + \epsilon\varsigma + o(\epsilon).$$

Then

$$L(x, \lambda, \epsilon) = L(x, \lambda) + \epsilon(\xi(x) + \langle \lambda, \eta(x) \rangle + \langle \varsigma, g(x) \rangle) + o(\epsilon).$$

Taking the partial derivative with respect to x^i and using the initial critical point condition, we find the equations in Theorem. □

11.2.2 All ingredients are evolutive

Initially, to an optimization problem, we attach the Lagrange function

$$L(x, \lambda) = f(x) + \langle \lambda, g(x) \rangle$$

and the critical point conditions

$$\frac{\partial f}{\partial x^i} + \langle \lambda, \frac{\partial g}{\partial x^i} \rangle \ = 0, \ i = 1, ..., n.$$

We consider a differentiable variation $x(\epsilon)$ and we introduce the general perturbation

$$L(x(\epsilon),\lambda(\epsilon),\epsilon)=f(x(\epsilon),\epsilon)+\langle\lambda(\epsilon),g(x(\epsilon),\epsilon)\rangle$$

together with the critical point condition

$$\left(\frac{\partial f}{\partial x^i}+\langle\lambda,\frac{\partial g}{\partial x^i}\rangle\right)(x(\epsilon),\lambda(\epsilon),\epsilon)\ =0.$$

By derivation with respect to ϵ, setting $\epsilon=0$, and denoting $\frac{dx^i}{d\epsilon}|_{\epsilon=0}=y$, we produce a variational linear system associated to the system describing the catastrophe manifold. Of course, we use the total differential operator

$$\frac{d}{d\epsilon}=\frac{\partial}{\partial x^j}\frac{dx^j}{d\epsilon}+\frac{\partial}{\partial\epsilon},$$

together with the variational objects (hypothesis)

$$\frac{dx^j}{d\epsilon}|_{\epsilon=0}=y^j,\ \frac{\partial f}{\partial\epsilon}|_{\epsilon=0}=\xi(x),\ \frac{\partial g}{\partial\epsilon}|_{\epsilon=0}=\eta(x),\ \frac{\partial\lambda}{\partial\epsilon}|_{\epsilon=0}=\varsigma.$$

The differential operators $\frac{\partial}{\partial x^i}$ and $\frac{\partial}{\partial\epsilon}$ commute.

Theorem 11.2.2. *At a critical point $x(\lambda)$, the tangent space to catastrophe manifold, if it exists, is described by the system*

$$\left(\frac{\partial^2 f}{\partial x^i\partial x^j}+\langle\lambda,\frac{\partial^2 g}{\partial x^i\partial x^j}\rangle\right)y^j+\frac{\partial\xi}{\partial x^i}+\langle\varsigma,\frac{\partial g}{\partial x^i}\rangle+\langle\lambda,\frac{\partial\eta}{\partial x^i}\rangle=0.$$

Proof. Applying $\frac{\partial}{\partial\epsilon}$ to the critical point condition, we find

$$\frac{\partial}{\partial x^i}\left(\frac{df}{d\epsilon}\right)+\langle\frac{\partial\lambda}{\partial\epsilon},\frac{\partial g}{\partial x^i}\rangle+\langle\lambda,\frac{\partial}{\partial x^i}\frac{dg}{d\epsilon}\rangle=0,$$

where

$$\frac{df}{d\epsilon}=\frac{\partial f}{\partial x^j}\frac{dx^j}{d\epsilon}+\frac{\partial f}{\partial\epsilon},\ \frac{dg}{d\epsilon}=\frac{\partial g}{\partial x^j}\frac{dx^j}{d\epsilon}+\frac{\partial g}{\partial\epsilon}.$$

Setting $\epsilon=0$, we get the linear system in Theorem. More precisely, at a nonsingular critical point $x(\lambda)$ of the catastrophe manifold, we have a unique solution $y^j(x(\lambda))$, i.e., an n-dimensional vector of parameter λ; at a singular point $x(\lambda)$ of the catastrophe manifold, we have a linear variety of solutions, i.e., n-dimensional vectors depending on λ and p parameters, $y^j(x(\lambda);\alpha_1,...,\alpha_p)$. □

11.3 Evolution of an optimal control problem

The constituents of a single-time optimal control problem are: the independent variable (generally speaking, time) t, the initial time t_0, the terminal time t_f, the state vector $x(t)$, the control vector $u(t)$, a cost functional J, and a dynamic constraint (controlled ODE system) or an isoperimetric constraint.

Let us consider an abstract *optimal control problem* of the form: *minimize the continuous-time cost functional*

$$J(u(\cdot)) = \int_{t_0}^{t_f} L(t, x(t), u(t))dt,$$

subject to the first-order dynamic constraints (the state controlled ODE)

$$\dot{x}(t) = X(t, x(t), u(t)).$$

The function L is called *Lagrangian.*

We consider an evolution over $\tau \in [0, \epsilon)$, where the variation (surface) $w = w(t, \tau)$, $\tau \in [0, \epsilon)$, is fixed by the partial flow

$$w_\tau = \varphi(t, \tau, w(t, \tau), v(t, \tau)),\ w(t, 0) = x(t),\ v(t, 0) = u(t).$$

The objective functional is changed into

$$\mathcal{J} = \int_{t_0}^{t_f} \mathcal{L}(t, \tau, w(t, \tau), v(t, \tau))dt,$$

with the initial condition

$$\mathcal{L}(t, 0, w(t, 0), v(t, 0)) = L(t, x(t), u(t)).$$

The ODE constraint is changed into a dynamic PDE

$$w_t(t, \tau) = \mathcal{X}(t, \tau, w(t, \tau), v(t, \tau)),$$

with the properties

$$\mathcal{X}(t, \tau, w(t, \tau), v(t, \tau))|_{\tau=0} = X(t, x(t), u(t)),\ \varphi_t = \mathcal{X}_\tau + \mathcal{X}_w \varphi + \mathcal{X}_v v_\tau.$$

For the new problem we can use the Hamiltonian

$$\mathcal{H} = \mathcal{L}(t, \tau, w(t, \tau), v(t, \tau)) + q_i(t)\mathcal{X}^i(t, \tau, w(t, \tau), v(t, \tau)),\ i = 1, ..., n.$$

Theorem 11.3.1. *The evolution of an optimal control problem is characterized by Euler-Lagrange PDE for*

$$\mathcal{L}_\tau = \frac{\partial \mathcal{L}}{\partial \tau} + \frac{\partial \mathcal{L}}{\partial w}\frac{\partial w}{\partial \tau} + \frac{\partial \mathcal{L}}{\partial v}\frac{\partial v}{\partial \tau} = \frac{\partial \mathcal{L}}{\partial \tau} + \frac{\partial \mathcal{L}}{\partial w}\varphi + \frac{\partial \mathcal{L}}{\partial v}\frac{\partial v}{\partial \tau}.$$

11.3.1 Evolution of a curve described by a second order ODE

ODEs arise in many contexts of mathematics and social and natural sciences. Specific mathematical fields include analysis, geometry and analytical mechanics. Scientific fields include much of physics and astronomy (celestial mechanics), meteorology (weather modelling), chemistry (reaction rates), biology (infectious diseases, genetic variation), ecology and population modelling (population competition), economics (stock trends, interest rates and the market equilibrium price changes).

Let us consider a C^2 curve $x = x(t)$, $t \in I$, $x(t) \in \mathbb{R}^n$, solution of the second order ODE

$$F(t, x(t), \dot{x}(t), \ddot{x}(t)) = 0,\ t \in J \subset I.$$

Its variation $w = w(t,\tau)$, $\tau \in [0,\epsilon)$, solution of the Cauchy problem $w_\tau = \varphi(t,\tau,w(t,\tau))$, $w(t,\tau)_{\tau=0} = x(t)$, is generally supposed to be a surface, solution of the PDE

$$\mathcal{F}(t,\tau,w(t,\tau),w_t(t,\tau),w_{tt}(t,\tau)) = 0,\ \ \mathcal{F}_{\tau=0} = F.$$

The new function $\mathcal{F}$ is solution of a linear PDE,

$$\frac{\partial \mathcal{F}}{\partial \tau} + \frac{\partial \mathcal{F}}{\partial w}\, w_\tau + \frac{\partial \mathcal{F}}{\partial w_t}\, w_{\tau t} + \frac{\partial \mathcal{F}}{\partial w_{tt}}\, w_{\tau tt} = 0,$$

since the derivatives $w_{\tau t}, w_{\tau tt}$ are obtained from evolution condition.

Remark 11.3.1. *This theory can be applied to auto-parallel curves,*

$$\ddot{x}^i(t) + \Gamma^i_{jk}(x(t))\dot{x}^j(t)\dot{x}^k(t) = 0,\ \ x(t_0) = x_0,\ \ \dot{x}^i(t_0) = \xi^i_{x_0}.$$

11.3.2 Evolution of a surface described by a second order PDE

We start with a C^2 function $u(x,y)$, $(x,y) \in D \subset \mathbb{R}^2$, solution of second order PDE

$$F(x,y,u(x,y),u_x,u_y,u_{xx},u_{xy},u_{yy}) = 0,\ (x,y) \in E \subset D.$$

The solution determines the surface $u = u(x, y)$. Its variation $w = w(x, y, t)$, $t \in [0, \epsilon)$ is solution of the Cauchy problem

$$w_t = \varphi(t, x, y, w(x, y, t)), \ w(x, y, t)|_{t=0} = u(x, y),$$

satisfying a new PDE

$$\mathcal{F}(t, x, y, w(x, y, t), w_x, w_y, w_{xx}, w_{xy}, w_{yy}) = 0, \ \mathcal{F}_{t=0} = F.$$

The function $\mathcal{F}$ is solution of a linear PDE,

$$\frac{\partial \mathcal{F}}{\partial t} + \frac{\partial \mathcal{F}}{\partial w} w_t + \frac{\partial \mathcal{F}}{\partial w_x} w_{xt} + \frac{\partial \mathcal{F}}{\partial w_y} w_{yt} + \frac{\partial \mathcal{F}}{\partial w_{xx}} w_{xxt} + \frac{\partial \mathcal{F}}{\partial w_{xy}} w_{xyt} + \frac{\partial \mathcal{F}}{\partial w_{yy}} w_{yyt} = 0,$$

since the derivatives $w_{xt}, ..., w_{yyt}$ are obtained from evolution condition.

Example 11.3.2. Let us take the Laplace PDE $u_{xx} + u_{yy} = 0$, with a solution (harmonic function) $u(x, y) = x^2 - y^2$. We impose the evolution $w_t = w + x^2$. It follows the Poisson PDE $w_{xx} + w_{yy} = 2e^t - 2$, as equation $\mathcal{F} = 0$. The function $w(x, y, t) = (2e^t - 1)x^2 - e^t y^2$ verifies both the evolution condition and the Poisson PDE. The uniqueness is connected to problems on PDEs with unique solutions.

Example 11.3.3. Now we start with wave PDE $u_{xx} - u_{yy} = 0$, and the solution $u(x, y) = x^2 + y^2$. We impose the evolution $w_t = w + x^2$. It follows the nonhomogeneous wave PDE $w_{xx} - w_{yy} = 2e^t - 2$, as equation $\mathcal{F} = 0$. The general solution of the last PDE is

$$w(x, y, t) = f(x + y, t) + g(x - y, t) + e^t x^2 - y^2$$

and the function that verifies both the initial condition and the evolution relation is

$$w(x, y, t) = (2x^2 + y^2)e^t - x^2.$$

Example 11.3.4. Let us consider the Tzitzeica PDE

$$z_{xx} z_{yy} - z_{xy}^2 = c(xz_x + yz_y - z)^4,$$

and a solution $z = \frac{a}{xy}$, $xy \neq 0$, $9a^2 = \frac{1}{c}$. We impose the evolution $w_t = w + x^2$, $w|_{t=0} = z$. It follows the evolution PDE

$$((w_{xx} - 2)e^{-t} + 2)w_{yy}e^{-t} - w_{xy}^2 e^{-2t} = c((xw_x + yw_y - w)e^{-t} + x^2(1 - e^{-t}))^4,$$

as equation $\mathcal{F} = 0$. The function $w(x, y, t) = \left(\frac{a}{xy} + x^2\right) e^t - x^2$ verifies both the evolution condition and the equation $\mathcal{F} = 0$.

Example 11.3.5. Let us take into discussion the Monge-Ampere PDE

$$z_{xx}z_{yy} - z_{xy}^2 = H(x, y, z, z_x, z_y),$$

and a solution $z = \varsigma(x, y)$. We impose the evolution $w_t = w + x^2$, $w(x, y, t)|_{t=0} = \varsigma(x, y)$. It follows the evolution PDE

$$((w_{xx} - 2)e^{-t} + 2)w_{yy}e^{-t} - w_{xy}^2 e^{-2t} = \mathcal{H}(t, x, y, w, w_x, w_y),$$

as equation $\mathcal{F} = 0$. The function $w(x, y, t) = \left(\varsigma(x, y) + x^2\right) e^t - x^2$ verifies both the evolution condition and the equation $\mathcal{F} = 0$.

Remark 11.3.2. *Evolution PDE - based methods are widely used in image processing and computer vision. For many of these evolution PDEs, we can formulate the following questions: (i) how are they created? (ii) what can we say about the existence and regularity of solutions? (iii) how to effectively implement them to produce the desired effects?*

In this Chapter, we studied the generation of an arbitrary evolution PDE starting from a stationary ODE or PDE.

On the other hand, PDEs can be regarded as evolution equations on an infinite dimensional state space. The solution $u(x, t)$ belongs to a function space in x at each instant of time t. For example, second order PDEs are evolution equations for second order ODEs.

The monographs [1] offer the reader a treatment of the theory of evolution PDEs with nonstandard growth conditions. This class includes parabolic and hyperbolic equations with variable or anisotropic nonlinear structure. Similar problems are found in [4].

Problem 106. *Given a second order PDE regarded as evolutionary equation,*

$$\mathcal{F}(t, \tau, w(t, \tau), w_t, w_\tau, w_{tt}, w_{t\tau}, w_{\tau\tau}) = 0, \ (t, \tau) \in E \subset D$$

with respect to the parameter τ, find an originating ODE

$$F(t, u(t), \dot{u}(t), \ddot{u}(t)) = 0, \, t \in J \subset I,$$

where $F = \mathcal{F}_{\tau=0}$, $u(t) = w(t, 0)$.

Example 11.3.6. Let us take the Laplace PDE

$$u_{xx} + u_{yy} = 0.$$

We interpret it as an evolution equation with respect to the parameter y. Since the general solution of the Laplace PDE is of the form $u(x,y) = f(x+iy)+g(x-iy)$, the attached generating ODE must be of the form

$$u''(x) = \varphi(x) + \psi(x), \text{where } u(x) = u(x,0) = f(x) + g(x).$$

Let us consider the wave PDE

$$a^2 u_{xx} - u_{yy} = 0$$

as evolution equation with respect to the parameter y. Since the general solution of the wave PDE is of the form $u(x,y) = h(x+ay) + k(x-ay)$, the attached generating ODE must be of the form

$$u''(x) = \sigma(x) + \eta(x), \text{where } u(x) = u(x,0) = h(x) + k(x).$$

Example 11.3.7. A conditioning system of two diffusion PDEs

$$w(x,y,\tau = (\tau^1,\tau^2)); \ w_{\tau^1} = w_{xx}, w_{\tau^2} = w_{yy}, \ w_{\tau^1} + w_{\tau^2} = 0,$$

on $Oxy\tau^1\tau^2$, has as trace on the plane Oxy, the Laplace PDE

$$w_{xx} + w_{yy} = 0.$$

Example 11.3.8. Let us consider two Newton Laws

$$u_{xx}(x,y) = F(x,y,u), \ u_{yy}(x,y) = G(x,y,u).$$

The equilibrium condition

$$F(x,y,u) + G(x,y,u) = 0$$

leads to Laplace PDE $u_{xx} + u_{yy} = 0$. Another condition, as for example,

$$F(x,y,u) = c^2 G(x,y,u)$$

gives the wave PDE $u_{xx} - c^2 u_{yy} = 0$.

11.3.3 Evolution of a curve satisfying a first order PDE

Problem 107. *Solve the first order pseudo-linear PDE*

$$a(x,y,u)\frac{\partial u}{\partial x} + b(x,y,u)\frac{\partial u}{\partial y} = f(x,y,u),$$

with respect to the unknown function $u(x, y)$, if the function $u(x, 0)$ is a solution of the Cauchy problem

$$u_x = g(x, u),\ u(x_0, 0) = u_0.$$

Solution. Let us show that this problem lies with a problem of evolution. We look for the evolution $u(x, y)$ of the curve $u = u(x)$ satisfying the Cauchy problem

$$u_x = g(x, u),\ u(x_0) = u_0,$$

with respect to the parameter y. We accept that the evolution is via the surface $w = w(x, y)$ described by first order pseudo-linear PDE

$$a(x, y, w)\,\frac{\partial w}{\partial x} + b(x, y, w)\,\frac{\partial w}{\partial y} = f(x, y, w),\ w(x, 0) = u(x).$$

To solve the problem, suppose we have two first integrals C_1, C_2 for the characteristic differential system

$$\frac{dx}{a(x, y, w)} = \frac{dy}{b(x, y, w)} = \frac{dw}{f(x, y, w)},$$

associated to the last PDE. The general solution of the PDE is given implicitly by

$$F(C_1(x, y, w), C_2(x, y, w)) = 0,$$

where F is an arbitrary function of class C^1 that we must determine. The function F can be recovered in three steps: (i) we replace $y = 0$, obtaining

$$F(C_1(x, 0, u), C_2(x, 0, u)) = 0;$$

(ii) taking the derivative with respect to x, it follows

$$\frac{\partial F}{\partial C_1}\left(\frac{\partial C_1}{\partial x} + \frac{\partial C_1}{\partial u}\,u_x\right) + \frac{\partial F}{\partial C_2}\left(\frac{\partial C_2}{\partial x} + \frac{\partial C_2}{\partial u}\,u_x\right) = 0$$

and hence

$$\frac{\partial F}{\partial C_1}\left(\frac{\partial C_1}{\partial x} + \frac{\partial C_1}{\partial u}\,g(x, u)\right) + \frac{\partial F}{\partial C_2}\left(\frac{\partial C_2}{\partial x} + \frac{\partial C_2}{\partial u}\,g(x, u)\right) = 0;$$

(iii) from the algebraic system $C_1(x, 0, u) = C_1,\ C_2(x, 0, u) = C_2$, we obtain the functions $x = x(C_1, C_2)$ and $u = u(C_1, C_2)$, which fix the relation in (ii) as a linear PDE

$$A(C_1, C_2)\frac{\partial F}{\partial C_1} + B(C_1, C_2)\frac{\partial F}{\partial C_2} = 0.$$

11.4 Self-evaluation problems

Problem 108. *Study different evolutions of the following optimization problems: (1) find* $\min_x f(x)$*, where* $f(x) = ax + \frac{b}{x}$*; (2) determine the minimum and maximum points of the function* $f : \mathbb{R} \to \mathbb{R}$*,* $f(x) = a\cos x + b\sin x$*,* $a^2 + b^2 > 0$*.*

Hint See any Calculus textbook (for example, [6], pp. 31-32).

Problem 109. *Verify that the extrema of the function* $f : \mathbb{R}^3 \to \mathbb{R}$*,* $f(x, y, z) = xyz$ *subject to the constraints* $x^2 + y^2 + z^2 = 1$ *and* $x + y + z = 0$ *are* $f_{min} = -\frac{1}{3\sqrt{6}}$ *and* $f_{max} = \frac{1}{3\sqrt{6}}$*. Study different evolutions of this problem.*

Problem 110. *Study different evolutions of the following optimization problems: (1) Find the minimum of Rosenbrock function* $f(x, y) = (1 - x)^2 + 100(y - x^2)^2$ *subject to* $(x - 1)^3 - y + 1 \leq 0, x + y - 2 \leq 0$*; (2) Find the minimum of Rosenbrock function* $f(x, y) = (1 - x)^2 + 100(y - x^2)^2$ *subject to* $x^2 + y^2 \leq 2$ *(constrained to interior of a disk); (3) Find the maximum of the entropy* $H = -\sum_{i=1}^{n} p_i \lg_2 p_i$ *subject to* $0 < p_i \leq 1$*,* $\sum_{i=1}^{n} p_i = 1$*; (4) Find the minimum of the function* $f(x) = c_1 x_1^{a_1} + ... + c_n x_n^{a_n}$ *subject to* $x_i > 0, x_1 \cdots x_n \leq P$*.*

Hint (1) global minimum $f(1, 1) = 0$; (2) global minimum $f(1, 1) = 0$; (3) global maximum $H(\frac{1}{n}, ..., \frac{1}{n}) = \lg_2 n$; (4) see [6], pg. 94.

Problem 111. *Study different evolutions of the following optimization problems: (1)* $\min u(x, y)$ *subject to Euler-Tricomi PDE* $u_{xx} + xu_{yy} = 0$ *and* $x^2 + y^2 \leq 1$*; (2)* $\min u(x, t)$ *subject to Buckmaster PDE* $u_t = (u^4)_{xx} + \lambda(u^3)_x$ *and* $x + 3t \leq 1$*.*

Problem 112. *Study different evolutions of the following optimal control problems: (1) Find the maximum of the functional*

$$I(u(\cdot)) = -\int_0^1 (x(t) + u^2(t))dt$$

subject to

$$\dot{x}(t) = u(t), \ x(0) = 0, \ x(1) = x_1;$$

(2) Resource allocation: production rate $x(t) \in [0, \infty)$*, portion of* x *reinvested* $u(t) \in [0, 1]$*, portion of* x *stored* $1 - u(t)$*, change of production rate* $\gamma u(t)x(t)$*,* $\gamma > 0$*, amount of stored profit* $[1 - u(t)]x(t)$ *and*

maximization of stored profit

$$\max_{u:[0,t_f]\to[0,1]} \int_0^{t_f} [1-u(t)]x(t)dt$$

subject to

$$\dot{x}(t) = \gamma u(t)x(t), \; x(0) = x_0 > 0.$$

Hint (1) The solution is

$$x^*(t) = \begin{cases} -2t^2e^{2t} & \text{for} \quad t \in \left[0, \frac{1}{2}\right] \quad \text{and} \quad u^* = -1 \\ e^{2t}(2t^2-1) & \text{for} \quad t \in \left(\frac{1}{2}, 1\right] \quad \text{and} \quad u^* = 1. \end{cases}$$

(2) Hamiltonian $H = (1-u)x + \lambda\gamma ux$, adjoint equation $\dot{\lambda}(t) = 1 - u^*(t) - \lambda(t)\gamma u^*(t)$, $\lambda(t_f) = 0$; optimal control $u^*(t)$ is 1 when $t \in [0, t_f - 1/\gamma]$ and $u^*(t)$ is 0 when $t \in [t_f - 1/\gamma, t_f]$. It is optimal to reinvest at the beginning.

Problem 113. *Newton's aerodynamical problem consists in determining the minimum resistance profile of a body of revolution moving at constant speed in a rare medium of equally spaced particles that do not interact with each other. Collisions with the body are assumed to be perfectly elastic. Problem: find*

$$\min_y R(y(\cdot)) = \int_0^r \frac{x}{1+y'(x)^2}\, dx,$$

where $0 \leq x \leq r$, $y(0) = 0$, $y(r) = H$, *and* $y'(x) \geq 0$. *Study different evolutions of this optimum problem.*

Hint The solution is given in parametric form: $x(u) = 2\lambda u, y(u) = 0, u \in [0,1]$; $x(u) = \frac{\lambda}{2}\left(\frac{1}{u} + 2u + u^3\right)$, $y(u) = \frac{\lambda}{2}\left(-\lg u + u^2 + \frac{3}{4}u^4\right) - \frac{7\lambda}{8}$, $u \in [1, u_{max}]$, where λ and u_{max} are solutions of the equations $x(u_{max}) = r$ and $y(u_{max}) = H$.

Problem 114. *Three classical examples of (linear) evolution equations are* ($t \in \mathbb{R}$, $x = (x^1, ..., x^n) \in \mathbb{R}^n$):

(1) Heat equation: $\partial_t u - \Delta_x u = 0$, *where the unknown is a function* $u : \mathbb{R} \times \mathbb{R}^n \to \mathbb{R}$. *Here, the initial data is the value of* $u(t,x)$ *at* $t = 0$, *i.e.,* $u(0,x) = u_0(x)$, $x \in \mathbb{R}^n$.

(2) Schrödinger equation: $i\, \partial_t u + \Delta_x u = 0$, *where the unknown is* $u : \mathbb{R} \times \mathbb{R}^n \to \mathbb{C}$. *The initial data is again the value of* $u(t,x)$ *at* $t = 0$, *i.e.,* $u(0,\cdot) = u_0(\cdot) : \mathbb{R}^n \to \mathbb{C}$.

(3) Wave equation: $-\partial_t^2 u + \Delta_x u = 0$, *where the unknown is* $u : \mathbb{R} \times \mathbb{R}^n \to \mathbb{R}$. *Here, we require initial values for both* u *and* $\partial_t u$, *i.e.,* $u(t,x)|_{t=0} = u_0(x)$ *and* $\partial_t u(t,x)|_{t=0} = u_1(x)$.

Change the parameter $t \in \mathbb{R}$ *into* $t = (t^1, ..., t^k) \in \mathbb{R}^k$ *(instead of time evolution we want a multitime evolution), and give explanations.*

Solution. We will take the multitime $t = (t^1, ..., t^k) \in \mathbb{R}^k$, were $\mathbb{R}^k$ is endowed with the product order. We consider a distinguished vector field $H = (h^\alpha), h^\alpha = h^\alpha(t,x)$, $\alpha = 1, ..., k$, on $\mathbb{R}^k$, for each $x \in \mathbb{R}^n$, and it will define the multitime derivation operator (along the direction H) $D_H u = h^\alpha u_{t^\alpha}$. Using this operator, we build the multitime PDEs (1) $h^\alpha u_{t^\alpha} - \Delta_x u = 0$, (2) $ih^\alpha u_{t^\alpha} + \Delta_x u = 0$. The existence of the solutions [5]: There exists an infinity of distinguished vector fields $H = (h^\alpha)$ such that a solution to the single-time linear heat equation or Schrödinger equation is also a solution to the multitime linear heat equation or Schrödinger equation.

(3) Starting from the single-time wave equation, we introduce the multitime $t = (t^1, ..., t^k)$, a distinguished symmetric linear connection $\Gamma^\alpha_{\beta\gamma} = \Gamma^\alpha_{\beta\gamma}(t, x, u, u_{t^\lambda})$ and a distinguished fundamental symmetric contravariant tensor field $g = \left(g^{\alpha\beta}(t, x, u, u_{t^\lambda})\right)$ of constant signature (r, z, s), $r + z + s = k$. Using a C^2 function $u(t,x)$, we build the Hessian operator $(Hess_\Gamma u)_{\alpha\beta} = \frac{\partial^2 u}{\partial t^\alpha \partial t^\beta} - \Gamma^\gamma_{\alpha\beta}\frac{\partial u}{\partial t^\gamma}$ and its trace, called ultra-parabolic-hyperbolic operator,

$$\Box_{\Gamma,g} u = g^{\alpha\beta}(Hess_\Gamma u)_{\alpha\beta}.$$

We define a multitime PDE as [6]

$$\Box_{\Gamma,g} u + \Delta_x u = 0.$$

Bibliography

[1] S. Antontsev, S. Shmarev, *Evolution PDEs with Nonstandard Growth Conditions*, Springer, 2015.

[2] J. F. Bonnans, A. Shapiro, *Optimization problems with perturbations; a guided tour*, Siam Rev., 40, 2 (1998), pp. 228-264.

[3] O. Calin, C. Udriste, *Geometric Modeling in Probability and Statistics*, Springer, 2014.

[4] L. C. Evans, *Partial Differential Equations*, Graduate Texts in Mathematics 19, AMS, 1998.

[5] L. L. Petrescu, *Multitime Schrödinger spatial solitons*, U.P.B. Sci. Bull., A, 77, 4 (2015), pp. 3-8.

[6] L. G. Matei, C. Udriste, *Multitime Rayleigh Solitons*, arXiv:submit/0328970 [math-ph], 30 Sept 2011.

[7] O. C. Schnürer, *Geometric Evolution Equations*, Lecture notes for the Summer School of the International Research Training Group Arithmetic and Geometry, 02.09.2007-07.09.2007, in Alpbach (Tirol), Austria.

[8] P. A. Samuelson, *Foundations of Economic Analysis*, Harward University Press, Cambridge, 1947.

[9] C. Udriste, E. Tanasescu, Minima and maxima of real functions of real variables (in Romanian), Technical Editorial House, Bucharest, 1980.

[10] C. Udriste, *Convex Functions and Optimization methods on Riemannian Manifolds*, Kluwer Academic Publisheres, 1994.

[11] C. Udriste, *Geometric Dynamics*, Kluwer Academic Publishers, 2000.

[12] C. Udriste, I. Tevy, A. S. Rasheed, *Dynamic of evolutive optimization problems*, U.P.B. Sci. Bull., Series A, 80, 2 (2018), 23-32.

[13] C. Udriste, H. Bonnel, I. Tevy, A. S. Rasheed, *Bilevel Disjunctive Optimization on Affine Manifolds*, Chapter 7 in Ed. Jan Valdman, Optimization Algorithms - Examples, IntechOpen, 2018, pp. 115-136, http://dx.doi.org/10.5772/intechopen.75643

Chapter 12

TASKS WITH MAPLE

Motto:

"The counsellors, when they saw that our king kept always silent
Let him be to think whatever he pleased
And they began squabbling, asking themselves
Who is the wisest man in the whole world?"

George Coşbuc - *The Philosophers and the Ploughmen*

In this Chapter, we resume the theory of dynamical systems and differential geometry through applied works of computation, graphing and simulation through Maple facilities. Mathematical topics discussed include: Dynamical systems; Invariant sets, conservation and dissipation; The stability of equilibrium positions and Hopf's bifurcation; Gyroscopic forces and collision avoidance; Nambu dynamical systems; Geometric dynamics and wind theory; Motions of a curve; Motions of a surface; Riemannian metrics of evolution; Evolution of Tzitzeica hypersurfaces; Dynamic of evolutive optimization pro- blems. Maple is both a high level symbolic calculator and a program development environment with a particularly wide and powerful subroutine library. To mobilize readers, we used some encouraging advice including "Solomon's parables" (the son of David, king of Israel).

12.1 Report on "dynamical systems"

Purpose of the report: To illustrate Maple software capabilities in dynamical systems issues.

Mathematical background: see the Chapter "Dynamical systems".

Work area: Write the following commands in Maple, following their order. Describe and execute each command.

The laboratory report: copy of Maple inputs and outputs as well as explanations given by each student separately.

Topics: (1) Discover a chaotic trajectory of the Lorenz system. (2) Simulate a trajectory of the van der Pol oscillator in the complex variable.

Maple entries:

Lorenz system:

> with(DEtools);

> DEplot3d($\{(D(x))(t) = 10 * (y(t) - x(t)),\ (D(y))(t) = 24.74 * x(t) - y(t) - x(t) * z(t),\ (D(z))(t) = x(t) * y(t) - (8/3) * z(t)\}$, $\{x(t), y(t), z(t)\}, t = -.97..50, [[x(0) = 1, y(0) = 1, z(0) = 1]]$, scene = $[x(t), y(t), z(t)]$, stepsize = 0.1e-2, linecolour = $sin(t)$);

van der Pol oscillator in complex variable:

> with(DEtools)

> phaseportrait($[(D(x))(t) = u(t), (D(y))(t) = v(t), (D(u))(t) = 0.1e - 1 * (1 - x(t)^2 - y(t)^2) * u(t) - 2^2 * x(t) - (10 * (x(t)^2 + y(t)^2)) * x(t),\ (D(v))(t) = 0.1e - 1 * (1 - x(t)^2 - y(t)^2) * v(t) - 2^2 * y(t) - (10 * (x(t)^2 + y(t)^2)) * y(t)],\ [x(t), y(t), u(t), v(t)], t = -10..10, [[x(0) = 1, y(0) = -1,\ u(0) = -.5, v(0) = 1]], stepsize = 0.1e - 1, scene = [x(t), y(t)], linecolour = sin((1/2) * t * Pi)$);

Enter other Maple commands.

Parable: If you don't give up on your first try, you will surely succeed in the second.

Comment: Maple helps us to analyse, to explore, to visualize and solve mathematical problems with modern techniques.

12.2 Report on "Invariant sets, conservation and dissipation"

Purpose of the report: To illustrate Maple software capabilities in description of invariant sets, conservation and dissipation.

Mathematical background: see the Chapter "Invariant sets, conservation and dissipation".

Work area: Write the following commands in Maple, following their order. Describe and execute each command.

The laboratory report: copy of Maple inputs and outputs as well as explanations given by each student separately.

Topics:

Show that the Lorenz system admits an ellipsoidal trap region of the form

$$rx^2 + \sigma y^2 + \sigma(z - 2r)^2 \leq C.$$

Maple entries:

Restart Maple and load the plots package.

> restart: with(plots):

Define the right-hand side of the Lorenz system.

$$f := \sigma * (y - x); g := r * x - y - x * z; h := x * y - b * z;$$

Define the Lyapunov function.

$> V := r * x^2 + \sigma * y^2 + \sigma * (z - 2 * r)^2;$

Compute the differential $dV(L)$ along the Lorenz vector field L. A constant signum of dV highlights a trap since, for example, $dV(L) \leq 0$ means that the angle between ∇V and $(\dot{x}, \dot{y}, \dot{z})$ belongs to the interval $[\frac{\pi}{2}, \frac{3\pi}{2}]$, i.e., a field line of L starting in the interior of the closed set $V(x, y, z) = C$ cannot go outside the boundary.

$$dV(L) := simplify(diff(V, x) * f + diff(V, y) * g + diff(V, z) * h);$$

Plot the set $V(x, y, z) = C$. The ellipsoid is obtained for a small value of the constant C.

$> \sigma := 10 : b := 8/3 : r := 40 :$

$> W := x^2/(b * r) + y^2/(b * r^2) + (z - r)^2/(r^2) :$

> implicitplot3d($\{V = 61000, W = 1.0001\}, x = -30..30, y = -70..70, z = 0..160$);

Remark. In such problems we also need the signum of a function of several variables.

If the function $f(x, y, z)$ is a quadratic form of matrix A, then to establish the sign we use

> with(LinearAlgebra);

$A := ` < | > `(` <, > `(1, 8, 3), ` <, > `(-4, 5, 2), ` <, > `(6, 1, 0));$

> IsDefinite(A);

> IsDefinite(A, 'query' = 'positive_semidefinite');

> IsDefinite(A, 'query' = 'negative_semidefinite');

To obtain the sign of a function $f(x, y, z)$, which is not a quadratic form, we can use a trick through Maple commands.

> with(plots);

> implicitplot3d($f(x, y, z) = abs(f(x, y, z)), x = -2..2, y = -2..2, z = -2..2, grid = [13, 13, 13]$);

Enter other Maple commands.

Parable: The fear of the Lord is the beginning of wisdom. Fools despise wisdom and instruction.

Comment: Maple offers capabilities for numerical and symbolic calculations, visualizations, optimizations, etc.

12.3 Report on "Local stability of equilibrium points and Hopf bifurcation"

Purpose of the report: To illustrate Maple software capabilities in stability problems, bifurcations, and chaos.

Mathematical background: see the Chapter on "Stability of equilibrium positions and Hopf bifurcation".

Work area: Write the following commands in Maple, following their order. Describe and execute each command.

The laboratory report: copy of Maple inputs and outputs as well as explanations given by each student separately.

Topics:

Show that the equilibrium point $(0, 0, 0)$ of the Lorenz system

$$\dot{x}(t) = \sigma(y(t) - x(t)), \dot{y}(t) = rx(t) - y(t) - x(t)z(t), \dot{z}(t) = x(t)y(t) - bz(t)$$

is asymptotically stable using the Lyapunov function

$$V(x, y, z) = x^2 + \sigma y^2 + \sigma z^2.$$

Maple entries:

> restart;

Define the Lorenz vector field L:

> $f := sigma * (y - x); g := r * x - y - x * z; h := x * y - b * z;$

Define the Lyapunov function.

> $V := x^2 + sigma * y^2 + sigma * z^2;$

Compute the differential $dV(L)$ on trajectories and analyse its signum.

$> dV(L) := simplify(diff(V,x)*f+diff(V,y)*g+diff(V,z)*h);$

Enter other Maple commands.

Parable: A wise man shall hear and shall be wiser: and he that understandeth, shall possess governments.

Comment Maple offers capabilities for numerical and symbolic computations, visualizations, optimizations, etc. The user can combine different commands for detailed solutions.

12.4 Report on "Gyroscopic forces and collision avoidance"

Purpose of the report: To illustrate Maple software capabilities in movement problems and the collision avoidance between two moving bodies.

Mathematical background: see the Chapter on "Gyroscopic forces and avoidance the collision".

Work area: Write the following commands in Maple, following their order. Describe and execute each command.

The laboratory report: copy of Maple inputs and outputs as well as explanations given by each student separately.

Topics: The movement equation of a mass-spring system is

$$m\ddot{x}(t) + Kx(t) - Ku_0 sin(\omega t) = F_0 cos(\omega t).$$

Fixing $K = 100, m = 400, = 1, u_0 = 0.1, F_0 = 1$ and the initial conditions $x(0) = 0$, $\dot{x}(0) = 0$, find the solution $x(t)$ by Maple techniques.

Maple entries:

> restart;

> EM := $m*(diff(x(t),t,t)) = -K*x(t)+K*u[0]*sin(omega*t) + F0*cos(omega*t)$;

> Init1:=$x(0) = x0$;

> Init2:=$D(x)(0) = v0$;

> dsolve($\{EM, Init1, Init2\}, x(t)$);

> $x(t)$:=rhs(%):

> $x\_plot := subs([v0 = 0, x0 = 0, K = 100, m = 400, F0 = 1, omega = 1, u[0] = 0.1], x(t))$;

> plot($x\_plot, t = 0..30$);

Enter other Maple commands.

Parable: That grace may be added to thy head, and a chain of gold to thy neck.

Comment: Maple offers capabilities for numerical and symbolic calculations, visualizations, optimizations, etc, but does not "get in" for the uninspired.

12.5 Report on "Nambu dynamical systems"

Purpose of the report: To illustrate Maple software capabilities in problems of special dynamical systems.

Mathematical background: see the Chapter on "Nambu dynamical systems".

Work area: Write the following commands in Maple, following their order. Describe and execute each command.

The laboratory report: copy of Maple inputs and outputs as well as explanations given by each student separately.

Topics: (1) Arrange the Euler dynamical system

$$\dot{x} = ayz,\ \dot{y} = bzx,\ \dot{z} = cxy$$

as Nambu dynamical system. Try graphically solving through Maple procedures.

(2) Find the general form of a vector field whose orbits are $x^2 - yz = c_1, y^2 - zx = c_2$. Simulate a trajectory.

Maple entries: Select commands from ODEtools, PDEtools.

Parable: Cast in thy lot with us, let us all have one purse.

Comment: Maple offers capabilities for numerical and symbolic computations, visualizations, optimizations, etc.

12.6 Report on "Geometric dynamics and wind theory"

Purpose of the report: To illustrate Maple software capabilities in describing the geometric dynamics and winds.

Mathematical background: see the Chapter on "Geometric dynamics and wind theory".

Work area: Write the following commands in Maple, following their order. Describe and execute each command.

The laboratory report: copy of Maple inputs and outputs as well as explanations given by each student separately.

Topics: (1) Transform the Euler flow

$$\dot{x} = ayz,\ \dot{y} = bzx,\ \dot{z} = cxy$$

in a wind using the Euclidean metric. Simulate a current path and a wind path through Maple programs.

(2) Transform the Goodwin flow

$$\dot{x} = \frac{1}{1+z^n} - ax,\ \dot{y} = x - by,\ \dot{z} = y - cz$$

in a wind using the Euclidean metric. Simulate a current path and a wind path through Maple programs.

Maple entries:

We refer to the case (1).

> restart;

The gyroscopic force is zero:

> with(VectorCalculus); SetCoordinates('cartesian'$_x, y, z$);

> F := VectorField(' <, > '(ayz, bzx, cxy));

> $\nabla \times F$;

> Curl();

It rests only the potential force ∇f, where

> $f := \frac{1}{2}(X1^2 + X2^2 + X3^2)$;

The wind is described by the differential system

$\ddot{x}(t) = (b^2z^2(t) + c^2y^2(t))x(t), \ddot{y}(t) = (c^2x^2(t) + a^2z^2(t))y(t), \ddot{z}(t) = (a^2y^2(t) + b^2x^2(t))z(t)$.

Otherwise, the theory can be applied, introducing other Maple commands.

Parable: For their feet run to evil, and make haste to shed blood.

Comment: Maple offers capabilities for numerical and symbolic computations, visualizations, optimizations, etc.

12.7 Report on "Motions of curves"

Purpose of the report: To illustrate Maple software capabilities in animation problems (geometric motion) of a curve.

Mathematical background: see the Chapter on "Motions of a curve".

Work area: Write the following commands in Maple, following their order. Describe and execute each command.

The laboratory report: copy of Maple inputs and outputs as well as explanations given by each student separately.

Topics: (1) Move a curve in plane or in space, using a parameter of evolution. (2) Generate particular curves.

Maple entries:

> with(plots);

The following two examples show how the curves animation in the plane is realized.

> animate(plot, $[A*x^2+x, x=-4..4], A=-3..3$);

> animate(plot, $[A*x^2-A^2*x, x=-4..4], A=-3..3, trace=5, frames=50$);

The following example shows how the animation is performed in three dimensions.

> animate(plot3d, $[A*(x^2+y^2)+x+y, x=-3..3, y=-3..3], A=-2..2$, style = patchcontour);

Let's animate a circle and a hyperbola given by equations default cartesian. Generalization to three dimensions refers to command plots[implicitplot3d].

> animate(implicitplot, $[x^2+y^2-r*x=r^2, x=-3..3, y=-3..3], r=1..3$, scaling = constrained);

> animate(implicitplot, $[x^2+A*x*y-2*y^2=1, x=-2..2, y=-3..3], A=-2..2$, scaling = constrained);

An animation of type "generation of curves", along the curves parameterized with respect to time, curves in two and three dimensions, can be done as in the following example. This example refers to a sine wave and a damped sine wave, but the process goes for any curve.

> animate(plot, $[[sin(t), sin(t)*exp(-(1/5)*t^2)], t=0..x], x=0..6*Pi$, frames = 50); We now take the trigonometric or rational parameterized circle. Generative animation can be of the type

> animate(plot, $[[cos(t), sin(t), t=0..A]], A=0..2*Pi$, scaling = constrained, frames = 50);

> animate(plot, $[[(1-t^2)/(1+t^2), 2*t/(1+t^2), t=-10..A]], A=-10..10$, scaling = constrained, frames = 50); Similarly, for a curve in space parameterized by time, we can think of an animation of type

> opts := thickness = 5, numpoints = 100, color = black; animate(spacecurve, $[[cos(t), sin(t), (2+ln(A))*t], t=0..20, opts], A=1..2$);

Enter other Maple commands.

Parable: But a net is spread in vain before the eyes of them that have wings.

Comment: Using the facilities provided by Maple, we can immediately capture the technical knowledge in electronic format that combines computations, explanatory texts, graphics, images, sounds and diagrams.

12.8 Report on "Motions of surfaces"

Purpose of the report: To illustrate Maple software capabilities in problems of animation (geometric motion) of a surface.

Mathematical background: see the Chapter on "Motions of a surface".

Work area: Write the following commands in Maple, following their order. Describe and execute each command.

The laboratory report: copy of Maple inputs and outputs as well as explanations given by each student separately.

Topics: (1) Move a surface in space, using an evolution parameter. (2) Generate particular surfaces.

Maple entries:

> with(plots);

Generically, we refer to animations of surfaces.

> animate3d($cos(t*x)*sin(t*y), x=-Pi..Pi, y=-Pi..Pi, t=1..2$);

> animate3d($x*cos(t*y), x=1..3, y=2..4, t=1..4$, coords = spherical);

> animate3d($1.3^x*sin(t*y), x=-1..2*Pi, y=0..Pi, t=1..8$, coords = spherical);

> animate3d($sin(x)*cos(t*z), x=1..3, z=1/4..7/2, t=1..4$, coords = cylindrical);

> animate3d($[uv, t-u, v*cos(t*u)], u=2..4, v=1..3, t=1..4$);

$>$ animate3d($[x, y, 1.3^x * sin(u * y)], x = 1..3, y = 1..4, u = 1..2$, coords = spherical);

$>$ animate3d($[x * u, t * u, x * cos(t * u)], x = 1..3, t = 1..4, u = 2..4$, coords = cylindrical);

$>$ animate3d($cos(t * x) * sin(t * y), x = -Pi..Pi, y = -Pi..Pi, t = 1..2, color = cos(x * y)$);

Among other things we can think of surface generation animations (two-dimensional evolution, good for optimal bi-temporal control).

$>$ restart;with(plots): $a := [t, t^2]; X := t- > t; Y := t- > t^2; f := (u, v)- > sin(u * v); g := (u, v)- > cos(u); h := (u, v)- > cos(v)$;

$> F := proc(t)localB; B := Array(1..3)$;

$B[1] := plot3d([u, v, 0], u = 0..X(t), v = 0..Y(t)$, axes = boxed, scaling = constrained);

$B[2] := spacecurve([X(s), Y(s), 0], s = 0..t$, color = black, thickness = 2);

$B[3] := plot3d([1+f(u, v), g(u, v), h(u, v)], u = 0..X(t), v = 0..Y(t))$;

plots[display](B[1], B[2], B[3]) end proc;

$>$ animate($F, [t], t = 0..1$);

Another variant of generation.

$>$ restart: with(plots); $a := [t, t^2]$;

$>$ plot3d($[sin(u*v), cos(u), cos(v)], u = 0..1, v = 0..1$,axes=boxed);

$>$ display($seq(plot3d([[x, y, cos(x^2 - y^2)], [x, y, 0]], x = 0..a[1], y = 0..a[2], axes = boxed), t = 0..1, .1)$, insequence = true);

$>$ display($seq(plot3d([[sin(u*v), cos(u), cos(v)], [u, v, 0]], u = 0..a[1], v = 0..a[2], axes = boxed), t = 0..1, .1)$,insequence=true);

Enter other Maple commands.

Parable: So the wage of every covetous man destroy the souls of the possessors.

Comment: Maple's intuitive interface supports multiple forms of interaction, from "Clickable Math tools" to sophisticated programming languages.

12.9 Report on "Riemannian metrics of evolution"

Purpose of the report: To illustrate Maple software capabilities in problems with highlighting Riemannian metrics of evolution.

Mathematical background: see the Chapter on "Riemannian metrics of evolution".

Work area: Write the following commands in Maple, following their order. Describe and execute each command.

The laboratory report: copy of Maple inputs and outputs as well as explanations given by each student separately.

Topics: (1) On the Riemannian manifold $(\mathbb{R}, g(t,x))$, analyse the metric flow determined by the Hessian

$$\frac{\partial g}{\partial t}(t,x) = \frac{d^2 f}{dx^2}(x) - \frac{1}{2g(t,x)}\frac{\partial g}{\partial x}(t,x)\frac{df}{dx}(x).$$

Detailed properties for the function $f(x) = x^3$.

(2) On the Riemannian manifold $(\mathbb{R}, g(t,x))$, study the dynamics (wave) of the metric determined by the Hessian

$$\frac{\partial^2 g}{\partial t^2}(t,x) = \frac{d^2 f}{dx^2}(x) - \frac{1}{2g(t,x)}\frac{\partial g}{\partial x}(t,x)\frac{df}{dx}(x).$$

Detailed properties for the function $f(x) = x^2$.

Maple entries:

Firstly, the flow.

```
> with(PDEtools);
> PDE := g(t,x)*(diff(g(t,x),t))+1.5*x^2*(diff(g(t,x),x))-
6*x*g(t,x) = 0;
> ans := pdsolve(PDE);
> build(ans);
```

Alternative:

```
> PDE1 := g(t,x)*(diff(g(t,x),t))+1.5*x^2*(diff(g(t,x),x))-
6*x*g(t,x) = 0;
> pdsolve(PDE1, build);
> build(pdsolve(PDE1));
```

Now wave.

```
> with(PDEtools);
> PDE2 := diff(g(t,x),t,t)+x*(diff(g(t,x),x))/g(t,x)-2 = 0;
> build(pdsolve(PDE2));
```

Enter other Maple commands.

Parable: Nor consented to my counsel, but despised all my reproof.

Comment: Maple offers breathing, depth and performance in handling any math problem.

Bibliography

[1] V. I. Arnold, *Geometrical Methods in the Theory of Ordinary Differential Equations*, Springer-Verlag, New York, 1983.

[2] M. Axenides, *Non Hamiltonian Chaos from Nambu Dynamics of Surfaces*, arXiv:1109.0470v1 [nlin.CD] 2 Sep 2011.

[3] O. Bernard, J.-L. Gouze, *Global qualitative description of a class of nonlinear dynamical systems*, Artificial Intelligence, vol. 136, pp. 29-59, 2002.

[4] U. Helmke, J. B. Moore, *Optimization and Dynamical Systems*, 2nd Edition, University of Würzburg, Germany; Australian National University, Canberra, Australia; March 1996.

[5] A. Julius, A. Halasz, V. Kumar, G. Pappas, *Controlling biological systems: the lactose regulation system of Escherichia Coli*, in American Control Conference 2007.

[6] C.-G. Lefter, *Differential Equations and Dynamical Systems*, (in Romanian) Alexandru Myller Editorial House, Iassy, 2006.

[7] S. Lynch, Dynamical Systems with Applications using Maple Companion software for *Dynamical Systems with Applications using Maple, 2nd Edition*, Birkhäuser, 2009.

[8] A. Neishtadt, *Lectures on Dynamical Systems*, Mathematics Access Grid Instruction and Collaboration (MAGIC) consortium, Loughborough University, 2007.

[9] G. Teschl, *Ordinary Differential Equations and Dynamical Systems*, Viena University, Austria, 2007.

[10] C. Udriste, *Geometric Dynamics*, Mathematics and Its Applications, 513, Kluwer Academic Publishers, Dordrecht, Boston, London, 2000.

[11] C. Udriste, *Differential Geometry, Differential Equations* (in Romanian), Universitext 15, Geometry Balkan Press, Bucharest, 2002.

[12] C. Udriste, L. Matei, *Lagrange-Hamilton Theories* (in Romanian), Monographs and Textbooks 8, Geometry Balkan Press, Bucharest, 2008.

[13] I. Vrabie, *Differential Equations* (in Romanian), MATRIXROM Editorial House, Bucharest, 2000.

[14] E. T. Whittaker, *A Treatise on The Analytical Dynamics of Particles & Rigid Bodies*, Cambridge University Press, 1989.

INDEX

Chapter 6 **101**

Chapter 7 **123**